IF 6 WAS 9

PRAISE FOR *IF 6 WAS 9 AND OTHER ASSORTED NUMBER SONGS*

"What may appear on the surface to be a light "bathroom book" is actually an exhaustively researched music-nerd call-to-arms. Fascinating, addictive, occasionally provocative, frequently uproarious, completely irresistible."
—**James Jackson Toth**, Wooden Wand

"Ultimately, this book aspires to find a heightened cultural understanding through its numerical lens, as if the notes of each song were tea leaves at the bottom of a cup. Klein transcends the novelty of his idea's barstool origins, delving sincerely and significantly into the mythos of each of these charged numerals that pop up in such a surprising number of the songs we love, hate or have never heard of. The particularly poignant stretch of 14 through 18 delves into the weird and nuanced transition from childhood into adulthood, achieving the kind of broad cultural analysis and commentary that draws readers to Greil Marcus' work. Klein also touches upon songs with numbers in their lyrics—not solely in their titles—to capture the fine psychic gradations of how coming of age is conceived."
—**Chris Vitiello**, *Independent Weekly*

"I want to be reborn as a patron of the Lotus bar on Clinton Street in New York's Lower East Side when David Klein was there in 2006 coming up with the idea of Pop Music Numerology. Klein is a major pop music nerd, but he's also a brilliant essayist with a huge and irresistible pair of ears."
—**Leslie Dunton-Downer**, author of *The English Is Coming!* and co-author of the *Essential Shakespeare Handbook*

"Alone, the chapters are a quick, satisfying snack. Taken together, they tell the story of a few decades' worth of pop music, bouncing off one another in weird sideways coincidence. It's like finding a corner café where Edwin Starr, Patti Smith, and Johnny Cash all happen to be hunkered down in separate corner booths. And then going to the Laundromat next door to see Iggy Pop, Chuck Berry, and Neil Young minding their own business, counting out quarters for the drier."
—**Jeff Klingman**, *The L Magazine*

"You think you know a lot about popular music? David Klein may not ever reach infinity, but he makes you realize there's a happy humming universe of music out there that you've either forgotten or need to explore."
—**Gerry Hadden**, reporter for PRI's The World and author of *Never the Hope Itself: Love and Ghosts in Latin America and Haiti*

"For me the book is the perfect blend of a writer who takes music seriously—one doesn't possess that much knowledge about something without taking that something seriously—and who writes about music with a tremendous sense of humor and sense of history."
—**Rick Cornell**, *No Depression, Independent Weekly*

"I feel certain it will be a major addition to the Rock Library here."
—**Mitch Easter**, producer, Fidelitorium Recordings

IF 6 WAS 9

And Other Assorted Number Songs:
From the No. 1 Song in Heaven to the 99th Floor

David Klein

White River Press
Amherst, Massachusetts

© 2016 David Klein

All rights reserved. No part of this publication may be reproduced, stored in a retrieval system, or transmitted, in any form or by any means, electronic, mechanical, photocopying, recording, or otherwise, without the prior written permission of the publisher.

Cover art by Nathan Golub and interior art by JP Trostle. Author photography by Jeremy Lange.

Images used in photo illustrations on pages 6, 53 and 76 are from dreamstime.com, and the bathroom wall on page 20 is courtesy of Donald Taylor II, cc by 2.0

Published by White River Press

ISBN 978-1-887043-21-2

First printing: January 2016

Library of Congress cataloging-in-publication data

Klein, David, 1962-
If 6 was 9 and other assorted number songs : from the no. 1 song in heaven to the 99th floor / by David Klein.
pages cm
ISBN 978-1-887043-21-2 (pbk. : alk. paper)
1. Popular music–Miscellanea. 2. Numbers, Ordinal–Songs and music-History and criticism. 3. Symbolism of numbers.
I. Title. II. Title: If six was nine and other assorted number songs.
ML3470.K54 2015 782.42164--dc23
2015026766

For Alison, my No. 1
My aim is true

Foreword

We live in an age of Geek. If the Geeks haven't exactly inherited the Earth, they have at least become a magnificently ubiquitous presence in this modern age we live in. Modern life is a whiplash juxtaposition of new horizons being explored and almost instantly abandoned at such a breakneck speed that we are often left longing for something a little more real, a little more substantial and lasting. Hence, the modern-day vinyl renaissance. Music can be amassed so easily and cheaply that some still yearn for the old-timey feel of holding something permanent, or at least physically real, in our hands: an album cover with a real, preferably heavy slab of vinyl that can only be played by the physical act of a diamond stylus riding upon a groove.

Into that backdrop of our collective time enters David Klein's new magnum opus of sonic numerology. *If 6 Was 9* is his exhaustive and highly entertaining listing and ranking of songs with numbers in the title. It's all here, from the seemingly obvious (Big Star's "Thirteen" and Hendrix's "If 6 Was 9") to the deep digging obscure (Magnolia Electric Co.'s "Texas 71" and Sleater-Kinney's "Dance Song '97"). Bonus points in my book for the inclusion of my band's song "72" and even bigger bonus points for also including Dickie Goodman's "Convention '72," a 45 of which I bought as a second grader and still own.

When I first heard about *If 6 Was 9*, I told the author that this was "Geek Shit at its finest" and of course I meant that as a compliment. Informative and just a ton of fun, it's a book to read and enjoy, then set upon the bookcase, overstuffed with records old and new. A useful aid to countless mix tapes and playlists, and tonic for many a late-night drunken debate, *If 6 Was 9* starts a conversation that won't be finished anytime soon.

—Patterson Hood
(Drive-By Truckers)

Introduction

It was conceived, like so many things, over a beer. I had recently moved my family from the East Village to the Lower East Side—a tiny distance geographically, but a world away in New York City. Whenever time allowed, I'd swing around the block to Lotus, a no-frills corner bar on Clinton Street with big windows and an eclectic mix of locals. Some were on the way up, like Stef, an unremarkable girl from the neighborhood who would stop by to drop off her gig fliers and smoke a cigarette with Ivan the bartender (apparently she was signed to Interscope and performed around town under the name Lady Gaga). Some were on the way down, like a cohort of regulars known as Rafael the Failed Poet, Full-of-Shit Ken, and Bad Ronald.

I was somewhere in the middle, holding my own but in daily need of a respite from a new teaching job I wasn't enjoying. Plus, my wife and I had twin boy toddlers, and our apartment was growing smaller by the day. In a city known for its intensity, I found that the low-key Lotus suited me, and it became my local. You could find me there most weekdays after work. With a few pints in me—some schemes posited and shit talked—I would head home to relieve the babysitter, feeling the better for it.

Ivan wasn't just the bartender at Lotus but its de facto DJ and resident music freak/geek. We became fast friends, of course, fueled by feverish conversations and shared playlists passed back and forth. Lotus became our living room. Whatever struck our fancy—the mix I had made the night before, *Loveless*, Patton Oswalt's routine about the notorious movie producer Robert Evans—we'd let it rip. If some schlub working on a laptop was distracted by the mind-bending force of Kevin Shields's guitars, or the sound of a maniac bellowing about Diane Keaton soaking in a Jacuzzi full of apple juice, so be it. We even had shtick. One riff went, "You know what they say... when in doubt, play the greatest record ever made." And we would.[1]

Not that we were completely impervious to our surroundings. When Stephen Malkmus of Pavement came in and sat at a booth with a coffee on a quiet fall afternoon, we definitely gave a shit. But as much as we tweaked the mix with songs we considered top-grade obscure/cool/genius, the proto-slacker was all business. He gave no indication that he even unconsciously detected the presence of music—not a single bob of the chin to acknowledge the beat. We were a bit aggrieved. But it was on one of those near-empty late afternoons that we struck gold.

---
1 *The Stone Roses* (1989)

David Klein

Ivan and I were discussing the most recent mix I'd made for him, specifically a song called "88," by art pop chanteuse Anna Domino. It's a really gorgeous, haunting thing that a friend had put on a mix for me years ago. Ever since, I had zealously shared it with the uninitiated. So we had "88" cranking and we were loudly singing its praises. Feeling a bit grandiose, I was moved to say something like, "Seriously, Ivan, this's gotta be the greatest #88 song ever written." In retrospect, it was a rash remark. Had I paused for even a second, surely "Rocket 88" by Jackie Brenston & His Delta Cats (often called the first rock 'n' roll song) or "88 Lines About 44 Women" by the Nails would have sprung to mind, and I would have qualified the statement. But my mental apparatus was already in overdrive.

"'Strawberry Letter #23,'" I sputtered. "That's gotta own 23, right? I mean, could there be a better #23 song in existence? The Zombies' 'Care of Cell 44'?? That has to be the ultimate #44 song *of all time*."

"Ah, good point."

"'I'm Eighteen,' I said, taking a quick gulp. "Alice Cooper. *Obviously…*"

"'24 Hours,' he offered. "Joy Division. Obviously."

"19th Nervous Breakdown."

"96 Tears.'"

We were off and running. By end of dusk, I had filled up three or four slightly damp bar napkins with song titles.

So we had definitely stumbled upon a new genus of trivia; that much was clear. What I had yet to realize was that I had, providentially, found my subject. I was already writing a column called "Obscurer Than Thou" on a pal's music blog called Merry Swankster, singing the praises of the Volcano Suns, Willie "Loco" Alexander, and others whom I believed needed to be celebrated in the blogosphere. Focusing on the deeper recesses of my record collection allowed me write about things that mattered to me, but not as deeply as I wanted to go. My record collection is large, not infinite, and I was starting to get the idea that numbers would be my way to, if not actually *touch* the infinite, get damn close.

Why number songs? Songs with numbers in their titles are a genre unto themselves. Because of the singularity of numbers and how they work in our lives, each one—at least those from 1 through 99—seems to occupy its own little fiefdom in a way that letters, or even words do not. Numbers have their own logic; they don't need us to make them mean something. But songs that share a number, especially an offbeat one, seem to be united in a very specific way. After examining the major #17 songs out there, for example, you quickly figure out that all the best ones are about adolescent girls. Clearly 88 had a hell of a flavor. I knew 18 had a flavor, thanks to Alice Cooper. Now I wanted to know about the rest of them.

At the outset, the wisdom of this venture (originally named "rock numerology") was not obvious. Friends advised me that it would be easy in the low numbers, but when I got to like, 34, forget it. Shortly after I began writing my column, my fellow bloggers summed it up this way: "Numerology is our pal Dave's ill-advised quest to find the definitive song for every number from one to a hundred. The higher the digit, the lonelier the climb." And they did have a point. But it was too late. Damn the doubters, naysayers, and the torpedoes—this thing was on.

It didn't take me long to establish that song titles containing even the crookedest of numbers, like 34 and 73 and 46, could indeed be found. Thank you, Interwebz. So the question became, is there a *good* song, a damn good one, for every number up to

99? (I soon decided that 99, not 100, would be my target goal.) Can I find 99 songs capable of standing shoulder to shoulder with "96 Tears" and "Eight Days a Week" without embarrassing themselves? No weak links, no head-scratchers, no "well, it was 43 and that's the best I could do"? I believed it was possible. Clearly, it would be a major undertaking, but I sensed that my foot would find a rock as I skipped across the pond, and I wouldn't fall in.

Eventually, I became an expert in my chosen field, an authority—*the* authority. At first, though, I was a novice. My M.O. was simply to brainstorm names of songs from memory, search my various music collections, and start writing. This approach was swiftly proved inadequate in "The Case of If 6'1" Was 9."

I'd awarded Liz Phair top honors in the #6 category for her song "6' 1"," based partly on the neat segue it made following Iggy Pop's "Five Foot One." In response, someone posted the comment, "What about 'If 6 Was 9'?" referring to the absolutely killer Jimi Hendrix song from *Axis: Bold as Love* that I had somehow missed. How could it have happened? I had always loved that song, and, no offense to Liz, more passionately than "6' 1"." Suddenly I was out of Plato's cave and seeing just how much effort would be required to do justice to this thing. I would have to strap myself in and follow the numbers like a hellhound. I would have to slow down, dig deeper, and perform an essential task that I'd always tended to shirk: my homework. And I was willing to do all that, as long as I could hold onto Spinal Tap keyboardist Viv Savage's credo and "Have a good time, all the time."

Still, my epiphany did not immediately take root. For my #26 essay, for example, the musical yield was initially so paltry that the chapter's final thought was something like, "And that's all there are. You can even check." I now know this to be the number-song maven's version of the last words of John Sedgwick, a Union general killed by a Confederate sharpshooter shortly after declaring, "Why they couldn't hit an elephant at this distance."

My own aim has improved as I've followed the numbers. I've refined a process for finding songs that pretty much guarantees I won't bypass something on the level of "If 6 Was 9." But these pieces are necessarily incomplete. New number songs are being written all the time. And I've had to consciously exclude songs in the interest of clarity and concision. Besides, even with careful searching, I'm bound to miss things, whole genres even, and quite possibly the best #22 song in the world. Still, I've thought about number songs more than just about anybody on the planet, so that has to count for something.

Seven years have passed since that fateful beer at Lotus, a fact that I find sobering. Lotus itself is long gone, having folded in fits and starts, shuttered, and re-opened as a faux-Irish bar. An inscription near the entrance, in fancy cursive, proclaimed, *"Home to lively debate and raucous revelry."* In short, not the type of place where you could play *Pet Sounds* all the way through multiple times, if even once.

By late 2008—post-market crash, pre-Obama first term—we were gone too, lured to Chapel Hill, N.C., by a job offer we couldn't refuse. I never pictured myself living in the South, but in the months leading up to our departure, the prospect of warmer weather, a backyard, and a less hectic existence started looking good. This project traveled well. In the not-without-its-bumps transition to a non-New York state of mind, nestling back into the familiar confines of "the quest" was a balm. I distinctly remember being in the midst of Chapter 59 as 2009 dawned, sifting through "The 59th Street Bridge Song" arcana and wondering when I would start feeling groovy in my newly adopted state.

Somewhere along the line—no, wait, that's a lie—when I reached #71—I decided to self-publish a first volume covering the teeming numbers from 1 through 33. I wanted to get something out there, share this thing with the world. Doing so forced me to revisit the early chapters to bring them in line with the more methodically done later pieces; this was tougher than I thought it would be and was occasionally downright humbling. The writing was far bloggier than I'd remembered. (I really didn't hit my stride until 35—a situation that mirrored my actual life.) But returning to the scene of the crime and correcting a few egregious errors was immensely satisfying. Merely rectifying the "Christine Sixteen"-over-"Sweet Little Sixteen" debacle made the entire effort worthwhile.

Self-publishing is not a simple process, but it was educational. I did my due diligence on the slippery concept of "fair use," consulting with a copyright expert and then eliminating or paring back quotes accordingly. Still, it had its upside; I received a few wonderfully dry emails from Dean Wareham about quoting lines from Galaxie 500's "Fourth of July," in which he commiserated with me regarding record company intransigence. An associate in Sparks' Music Synchronization department was less inclined to dis the Man.

When it came out in softcover and digital form in early 2012, I discussed number songs on the local NPR affiliate show, "The State of Things," with Frank Stasio; I did a few readings at bookstores and got some nice reviews. Then it was back to work, and in fact, the hardest part of the journey. As with mountain climbing, the higher you go, the thinner the air.

The slow gestation period has been a blessing though. I've had eight years to stumble upon small but vital facts, to accrue listens and re-listens, to discover old obscurities as well as new number songs. I'm now convinced this project will never be done. As long as I'm able to listen to music, my ears will continue to be attuned to the numbers, and I'll keep on tinkering with this thing like it's an old car, maybe until the twelfth of never.

THE RULES

The definitive song must have the number in its title.

The number has to stand alone (so "1999" would not be eligible for 19 or 99; "In the Year 2525" is only eligible for top honors in the #2,525 category. Sorry, Zager & Evans fans).

Ordinals are OK ("Positively 4th Street"); so are apostrophized constructions ("Cruel Summer '89").

Classical music compositions (Mozart's Symphony No. 27 in G major K) are not eligible. It would be idiotic to try to compare the relative merits of Haydn's Symphony No. 96 with "96 Tears."

Rock and pop of the past 60 years and the roots thereof—blues, R&B, soul, country, rockabilly—are the critical genres under consideration. Songs from other modern genres, like jazz and show music, may be considered, but they will not take top honors.

… "One Fine Day," "One Too Many Mornings," "One More Cup of Coffee," "One For My Baby (and One More For the Road)," "One of These Days," "One of These Nights," "One More Saturday Night," "One Night in Bangkok," "One O'Clock Jump," "Just One of Those Things," "One Mint Julep," "One Headlight," "One Long Pair of Eyes," "One of Our Submarines is Missing," "One Nation Under a Groove," "One Tin Soldier (The Legend of Billy Jack)," "One After 909," "One Way or Another," "One Way Out," "One Bad Apple," "One Monkey Don't Stop No Show," "One on One," "One Love," "One of a Kind Love Affair," "Little Bitty Pretty One," "Long Cool One," "I'm One," "You're the One," "This Is the One," "I Am One," "Still the One," "She's the One," "Going for the One," "Could You Be the One?," "The One I Love," "The Only One I Know," "One of a Very Few of a Kind," "You're the One That I Want"…

That's how it starts—with the mother lode. No other numeral comes close. One has so many ramifications and crops up in so many critical figures of speech—in one-horse towns, on one-way streets—that it looms over its numerical brethren like the monolith in *2001*, which it kind of resembles. In fact, the field is so well stocked that choosing the definitive #1 song is a purely subjective act. There are just too many good ones. Significant whittling is called for, but that's my job.

It's simple though. By requiring the winning #1 song to have a pronounced sense of 1-ness (not oneness—that's different) we can pretty much dispense with the songs listed above, which are not about 1 or 1-ness, or even oneness. They concern a single headlight, a tin soldier, a night in Bangkok. Still, choices abound.

Songs called "Number One" make strange bedfellows of Joni Mitchell, John Legend, Styx, Pharrell, Hall & Oates, Deep Blue Something (a one-hit wonder if ever there was one), Etta James, Helloween, Martha Reeves, and my favorite "Number One," the libidinous thumper by Alison Goldfrapp. "Looking Out for Number 1" (which vied with "taking care of business" as the pet phrase of ego-driven '70s types) has been employed as a title by BTO, UFO, the 5th Dimension, and Travis Tritt.

Naturally, a flotilla of songs have been called simply "One" or "1," by a comically catholic assemblage that includes the Bee Gees, Busta Rhymes, Creed, Dokken, Vince Gill, Ghostface Killa, Alanis Morissette, and Sunny Day Real Estate. The "1" in "1-2-3 Red Light" by 1910 Fruitgum Co. represents a real 1, but it's no more important than 2 or 3. Manfred Mann's "5-4-3-2-1" and others of that ilk share this same basic shortcoming. "One," the mighty antiwar epic from Metallica, barely mentions the title number at all.

For me it all comes down to a trio of songs that wear #1 proudly on their sleeves. "One" is among the greatest songs in the U2 catalog—it could even be called definitive. The atmospheric slow build, crafted with Enovian precision, showcases the band's individual parts beautifully and leads to a truly joyful release. And the lyrics are sharp, powerful, and bombast-free. "One" is U2's most covered song. Johnny Cash, Mary J. Blige, and Joe Cocker are among the big voices to stand and deliver on "One"'s demands.

Then there's "One" (as in "the loneliest number"), a magical pop single written by Harry Nilsson, whose numerical expression of solitude is, well, singular. In 1968, Three Dog Night's version of Nilsson's "One" started off the band's run of 21 consecutive chart hits. Aimee Mann's take, prominently featured in *Magnolia*, strips away some of the falsetto bathos of previous versions. But my only real critique of "One" by U2, Aimee Mann or Three Dog Night is that they are not "The No. 1 Song in Heaven" by Sparks.

THE VERDICT

Shortly after I'd had my first glimpse of Sparks—on TV at soccer camp in 1975, performing "Reinforcements" on *Don Kirshner's Rock Concert*—I spent my meager savings on a copy of *Propaganda*. It was the first record I bought with my own money. For a long time I chose to withhold from my teenage friends how much I was both blown away and perversely fascinated by the prancing, falsetto-voiced, staccato-singing Russell Mael and his winsome, stock-still, Hitler-mustachioed, keyboard-playing brother, Ron. Songs like "Reinforcements," which they absolutely killed on *Kirshner*, were just so *stuffed*: stuffed with layers of fat glam guitars, stuffed with tasty words, like the made-up *subterfusion* and the might-as-well-be-made-up-for-all-the-French-I-knew *coup d'etat*. Most of all, Sparks were stuffed with an abundance of wit, smarts, and style.

And the rest of the record did not disappoint: there were more interesting words (*potentate, impetus, ornithologist*) and an abundance of astounding Les Paul hooks, not to mention the drum stylings of Norman "Dinky" Diamond, whose biopic beckons to be made. Thirty-five years after the record's release, I continue to prize the words of Jim O'Rourke, the multi-talented musician, producer, filmmaker, and occasional writer who called *Propaganda* "the standard to which I hold myself and everything else."

Sparks have had a go at nearly every musical idiom of the past three decades. The L.A. natives never tried grunge, but they aced the exam for lethal glam rock, orchestral bubblegum, and calibrated slabs of oomph like "The No. 1 Song in Heaven." When they decided to go the pure dance route, the Maels went straight to the top, enlisting the Eno of Disco himself, Giorgio Moroder. Not surprisingly, the entire platter pulsates like a finely crafted Dancing Machine delivered from on high. Beyond the personal experience of having had my life changed by Sparks, there is another perfectly logical reason to kick off this mad quest in this way. On the subject of 1, 1-ness or oneness, "The No. 1 Song in Heaven" is unimpeachable; it leverages the implications of "No. 1" as in "the pinnacle" and soars to celestial heights.

A SONG TITLE	AN ALBUM	A BAND	A LYRIC
"One of These Things First" – Nick Drake (1971)	#1 Record – Big Star (1972)	The Only Ones – London, punk/new wave (1978–1981)	"We are not two we are one" – The Kinks, "Strangers" (1970)

2

We live in a binary world: Adam and Eve, ones and zeroes, black and white, and the old standby, good old life and death. Even the basic mechanism of human existence on a cellular level—sodium in, potassium out—is a two-part sequence. Every Alpha needs its Omega. Love to the second power still inspires.

In the world of song, two often means one thing: two people in love. Many of the key two-titled songs deal with love or relationships, like Bill Withers' "Just the Two of Us," which only a curmudgeon could resist. The lovers in Hoagy Carmichael and Frank Loesser's "Two Sleepy People"—a foggy little fella and a drowsy little dame—are romance incarnate in their cozy chair, "pickin' on a wishbone from the Frigidaire." Pairs of hearts are plentiful: "Two Hearts" has been tapped as a title by Bruce Springsteen, Chris Isaak, Phil Collins, Vince Gill, John Hiatt, Pat Boone, Graham Nash, and the Jayhawks, among others, while Kylie Minogue, Digitalism, and Toto went with "2 Hearts." U2's "Two Hearts Beat as One" adds a little twist, and the list goes on and on. Of course, when a heart gets broke, it's always in two.

PALATE-CLEANSING #2 SONG FACT:
The most recorded #2 song in existence is "Tea For Two," a standard written by Irving Caesar and Vincent Youmans that has appeared on well over 1,500 records.

The Beatles' "Two of Us," whose fractious genesis was captured in the *Let It Be* film, deserves an exalted place on any list of great #2 songs. Beginning as a straight-ahead mid-tempo rocker, it only came together when stripped down to a rustic blend of voices, an emphatically strummed acoustic guitar, and a simple metronomic beat. A multitude, from the Beatles on down, have castigated Phil Spector for the grandiose liberties he took as producer of the *Let It Be* album, but "Two of Us," at least, is notably understated. Thus, the version on the un-Spectorized *Let It Be…Naked*, released in 2003, is pretty much like the original, only without Lennon's introductory "I dig a pygmy by Charles Hawtry and the Deaf Aids" bit (which has baffled and amused me since I was a child).

Plain-old "Two" has served as a song title many times, for everyone from Ryan Adams to Eyeless in Gaza, Porter Wagoner to Billy Squier, but none are as distinctive as songs titled "One," which is probably why none come quickly to mind. Here's a much richer vein; what I call the Two Nouns category. To wit: "Two Tribes" by Frankie Goes to Hollywood, "Two Heads" by the Jefferson Airplane, "Two Rooms" by the Feelies," "Two States" by Pavement, "Two Halves" by My Morning Jacket, "Two Gunslingers" by Tom Petty, "Two Trains" by Little Feat, "Two Hands" by King Crimson. I'm partial to "Two Sisters" by the Kinks, in which Ray Davies conjures the mutually jealous Sylvilla and Percilla in a few deft strokes. It culminates as one sister, realizing she has it better than her sibling, "ran around the house with her curlers on." And the song clocks in at 2:02. How 2 can you get? Also close to my heart is "Two Librans" by the Fall, a delectably dark and truculent seether that could almost pass for a Pixies song were it not for the inimitable slurred doggerel of Mark E. Smith. Only discrete snatches of lyrics are comprehensible, but he growls out the ostensible chorus, "Two Librans … *reflect*" with chilling precision.

Like Mr. Smith, Jim Morrison enjoyed a bit of the old grape. He was also fond of the ladies. "Love Me Two Times," a sassy single penned by Doors guitarist Robbie Krieger, was alleged to be a veiled reference to oral sex, but since when did the Doors veil anything? After the Oedipal freak-out of "The End" you'd think they would just come out and say it. More likely, the song depicts a soldier's plea to his beloved for an extra go-round before heading off to war. Also of note on the "two times" tip: Johnny Cash's "Two-Timin' Woman" and a slew of songs attesting to a cornucopia of treacherous two-timers: two-timin' mamas, papas, daddies, losers, babies, two-steppers, and turkeys.

Close kin of the Two Nouns category is the Two + Adjective + Noun category, with songs like "Two Fat Sisters" by the Clean (a riff on the Kinks song perhaps?), "Two Left Feet" by Richard Thompson, and "Two Fat Feet," a sexy two-chord vamp by Fiery Furnaces. You wouldn't necessarily think that a separate category for Two-Headed songs would be warranted, but you would be wrong: "Two Headed Man" is by bluesman Lonnie Brooks; "Two Headed Woman" is by bluesman Willie Dixon. "Debbie Gibson Is Pregnant With My Two-Headed Love Child" is the work of Mojo Nixon, and there are at least 20 more, with titles like "Two Headed Freap," "Two Headed Calf," and "Two Headed Alarm Clock." For me, the best representatives of the two-headed song genre make an appropriately bifurcated duo: Neutral Milk Hotel's heart-stopping "Two Headed Boy" and Roky Ericson's blistering "Two Headed Dog," which simply refuses to heel.

The Numeral 2 category is packed. There's 2 as in "to" (e.g., "Nothing Compares 2 U by Prince," "2 Kool 2B Forgotten" by Lucinda Williams); 2 as in "Part 2" (e.g., "Eye of Fatima Pt. 2" by Camper Van Beethoven); and 2 as in the second song on a record ("Song 2" by Blur). There's 2 as part of a sequence of numbers and/or letters, a rich subcategory packed with gems like "2HB" by Roxy Music, "V-2 Schneider" by David Bowie, "Y2K" by Apples in Stereo, "5-4-3-2 Wave" by the Patti Smith Group, "2-4-6-8 Motorway" by the Tom Robinson Band, and Radiohead's knotty, mathematically innovative "2 + 2 = 5."

Two has proven extremely useful in denoting chunks of time, from Ray Charles's "Two Years of Torture" to "Two Weeks" by FKA twigs. Defying easy categorization but, well, 2 kool 2b forgotten, are Mission of Burma's shimmery incantation "Trem Two," Magnetic Fields' "Two Characters in Search of a Country Song," which shimmers in a totally different way, Spoon's inscrutable "Two Sides of Monsieur Valentine," and the Pale Saints' "You Tear the World in Two." The ageless "Billy Two," a bit of offhand genius by The Clean, makes the most of feverish strumming, un-synced chorus vocals, and a surprise snatch of backward tape seemingly flown in from the fadeout of the Beatles' "Rain." And let's not forget Ted Weems, whose band played at the inauguration of Warren G. Harding and had a hit with "My Cutey's Due at Two-to-Two Today."

As with 1, so with 2—and most of the numbers you can count on one hand: the choices are vast. Some are flat-out great; many are very, very good, and lots have their cheesy/cringe-y charm.[1] It's a flood though. The best approach seems to be to break down the glut into basic categories and zero in on the most striking examples. Consider this an idiosyncratic survey accompanied by a diffidently offered choice of best #2 song ever. Applying the same standard as with 1, I'll insist that any song claiming this crown must address the essential character of the number. "Takes Two To Tango," for example, a lilting rhapsody

---

1 Like "Two Tickets to Paradise" by the rocker born Edward Mahoney, or "Two of Hearts" by Stacey Q, just name two.

written by Al Hoffman and Dick Manning and popularized by Louis Armstrong, Pearl Bailey, and Dean Martin, delineates the magic of two with irrefutable logic.

> *"You can croon to the moon by yourself*
> *Well you can laugh like a loon by yourself*
> *Spend a lot go to pot on your own*
> *There are a lot of things that you can do alone*
> *But it takes two to tango"*

Few would dispute that for most of the 20th century, this was the No. 1 #2 song. But three-plus decades after Armstrong and Bailey had hits with it, the Harlem duo of Rob Base and DJ EZ-Rock retooled tangoing for a new era with "It Takes Two." *Spin* magazine proclaimed it to be no less than the greatest single of all time. While any such declaration runs the risk of hyperbole, there's no denying the power of this sample-heavy joint, which lifts the primo vocal hooks from "Think (About it)," by soul chanteuse Lyn Collins, and has itself been liberally sampled.

THE VERDICT

Marvin Gaye's "It Takes Two," written by Motown songwriting aces Sylvia Moy and William "Mickey" Stevenson, is, to my ears, the most two-alicious #2 song out there. Echoing the lyrical sentiment of "Takes Two to Tango," this ebullient duet (it's a duet—what could be two-ier?) was an international hit in 1967 and captures the alchemy of two people in love in the alternating verses between Gaye and Kim Weston. Less celebrated than Tammi Terrell, Gaye's more famous collaborator, Weston more than holds her own here. "It Takes Two" dresses up classic odes to twosome-ness in crisp Motown duds, and it sparkles, undiminished, to this day.

A SONG TITLE	AN ALBUM	A BAND	A LYRIC
"Two Steps From the Blues" – Bobby "Blue" Bland (1961)	*Two Steps From the Middle Ages* – Game Theory (1988)	2Pac – West Coast, hip-hop (1990s)	"So me and you, just we two/ Got to search for something new" – Roxy Music, "Virginia Plain" (1972)

David Klein

3

Three really is the magic number. It's the digit that connects Moe, Larry, and Curly with the Father, the Son, and the Holy Ghost and every other immortal power trio, even the James Gang. The "ancient mystic trinity" from *Schoolhouse Rock!* shows up in exalted places across cultures, in science, art, religion, technology, and beyond. The very atoms that make up our world consist of three parts. Perhaps even more telling, without 3 there would be no funny. For the joke to work, you need all three of the Jew, the Irishman, *and* the Italian to walk into a bar. Nevertheless, 3 also has the darker connotation of the odd man (or woman) out: the fabled third wheel. Or as George Jones put it, "One is lonely, two's a marriage, three's a crowd." No doubt, #3 songs run the gamut from magical to miserable, but what they all share is the immediacy of the tripartite arrangement, something we perceive on a primal level.

It's hard to know what to make of the handful of #3 songs by bands or artists whose careers were tragically cut short, but the roster is pretty striking. Jimi Hendrix's cosmically groovy "Third Stone From the Sun" is endowed with the line "And you'll never hear surf music again" and was nicked nakedly by the shirtless Right Said Fred for "I'm Too Sexy." Nick Drake's quietly harrowing "Three Hours" limns the futility of escape. A triad of steps unite Lynyrd Skynyrd's "Gimme Three Steps," a good-natured stomp in which Ronnie Van Zandt pleads for a running start from a barroom badass, and "Three Steps to Heaven," Eddie Cochran's spooky, posthumous U.K. chart-topper in 1960.

Three is well represented among heavy hitters, sometimes in exalted fashion, sometimes not. The Beatles' cover of Leiber and Stoller's "Three Cool Cats" is a trifle from the pre-"Love Me Do" era. "Three Angels" is something of a curiosity in the Bob Dylan canon, a halting spoken-word piece with a slowly swelling gospel choir and lacking a verse-chorus structure. Judging from its placement as the penultimate song on *New Morning*, one suspects even Dylan might not have been thrilled with it. "Three Little Birds" is one of Bob Marley's most popular songs—certainly it's the happiest #3 song in existence. Standing in the way is a lack of 3-ness as well as overfamiliarity due to its being co-opted by the Jamaican tourist board.

Three p.m. is a dull time of day, but three in the morning is rife with potential. The KLF's "3 A.M. Eternal" speaks to the dance-floor moment when time disappears. Gary U.S. Bonds' jukebox gem "Quarter to Three" salutes that same lost-in-the-music feeling. In Johnny "Guitar" Watson's "Three Hours Past Midnight," the distraught axe-man bewails the lateness of the hour and the absence of his baby, and contemplates catching that midnight train. Guadalcanal Diary's "3 A.M." is a dark night of the alcoholic soul, while Matchbox Twenty's "3 A.M." chronicles the romantic yearnings of an insomniac who doesn't want her man to catch a cold. Then there's Young Jeezy's "3 A.M.," which is also about romantic yearnings, the kind that hit you when you're up on "that Grey Goose, higher than a pelican."

Jonathan Richman's youthful protagonist in "Not Yet Three" shares Young Jeezy's disdain for sleep, rejecting his bedtime with fierce and clear-eyed logic: "I'm stronger than you, you're simply bigger than me." Some rock 'n' roll purists poo-poo Richman's post-Modern Lovers output, but no one will ever write a more compelling song about the indignity of having to go to bed when you can still hear kids playing outside. Richman reminds us that the urge to rebel does not necessarily begin in the teen years. Mark Kozelek of Red House Painters is living proof of that. The winsome but wounded protagonist of his "Three-Legged Cat" would make a good companion for Richman's not-yet-3-year-old.

Several songs from the post-punk era are united by the number 3 and a certain astringency. Wire's "Three Girl Rhumba" is one of the easiest songs to sidle up to from the highly influential *Pink Flag*. It's a model of caustic, catchy precision, and the song's lyrics ("Open your eyes/Think of a number/Don't get sucked under/A number's a number") provide ample material for mathematical contemplation. Concurrently, the Cure released their debut album in the U.K., *Three Imaginary Boys*. The title track is characteristically stark, with watery strumming and a precise, plodding bass line over which Smith, his youthful croak full of foreboding, pleads, "Can you help me?" "Three Cheers for Our Side," by the semi-legendary Orange Juice, exemplifies the band's off-kilter charm, jaunty rhythms, and knack for the well-placed ooh-la-la.

Say what you will about the arrangement known as the threesome, but it has provided ample material for songwriters over the years. Jane's Addiction's "Three Days," a 10-minute opus from *Ritual de lo Habitual* with a "three lovers/in three ways" scenario, was apparently based on an actual incident. Nothing's shocking, especially when confessed by Perry Farrell. These days, when Britney Spears releases a song like "3"—an ode to the ménage-a-trois, sporting lyrics like "Countin' 1, 2, 3/Peter, Paul, and Mary"—the only people who find it disturbing are those prompted to imagine the archetypal '60s protest trio engaged in the song's controversial coupling … or would that be tripling? But at least Britney has fun in mind; Interpol's "No. 1 in Threesome," with its "teeth marks of time," is about as sexy as a Windows users' manual.

David Crosby might have been the first to broach the topic frankly with "Triad," in which he—with a lack of compunction characteristic of the "If it feels good, do it" school of thought—asked of his long-haired lovers, "Why can't we go on as three?"[1] No one would ever mistake the Commodores' "Three Times a Lady" for a song about a threesome, but in the hands of a more provocative singer, Peaches perhaps, these implications could be drawn to the surface. The Magnetic Fields' *Distortion* opens with the clangorous "Three-Way," which calls to mind both the Jesus & Mary Chain and Phil Spector but never gets into specifics (the only words are the joyously gang-vocaled title phrase). But our friend Jim O'Rourke fearlessly lets his freak flag fly in the deceptively pretty "Halfway to a Threeway," a hushed come-on to an apparently wheelchair-bound seducee: "I just can't get you to sit/ You and your stupid epileptic fits/And I know that you can't run away/ cause I'm halfway to a threeway."

A final word on the subject: In a recent survey, the number of American respondents who claimed to have participated in threesomes was bested only by the extremely tricurious Icelanders. (Norway captured the bronze.)

---

1 In 2015, I asked Mr. Crosby on Twitter whether the triad relationship could have worked under the right circumstances. He replied, "Did work but hard."

THE VERDICT

"Three Is a Magic Number" is perhaps the best-known song from *Schoolhouse Rock!*, the 1970s educational animated series that has since wormed its way into the pop cultural firmament. De La Soul's "The Magic Number" is firmly based on it, although the word "three" (or 3) does not appear in the title. Just this once, in the case of this audacious display of rhythm and rhyme and threeness, I'm going to ignore the title's obstinate shortcoming. The song's mathematical provenance and three-centric bent so fulfill my essence-of-the-number criterion that I am compelled to look the other way.

The De La Soul trio—Mase, Posdnuos, and Trugoy—nab the best parts of the original song by Bob Dorough (an accomplished jazz composer and pianist); they dispense with the times-tables practice and turn out a world-class jam that's not so much built on but crammed with a barrage of select samples. There's Eddie Murphy ("Anyone here ever get hit by a car?") and a shrieked James Brown count-off. Johnny Cash inquires "How high's the water, Mama?"; Fiorello LaGuardia reads the funnies ("What does it all mean?"); and Syl Johnson exhorts us to "*Do the shing-a-ling!*" All of it weaves in and out of some beats you'd have to be dead to ignore.

De La's 1989 calling card, *3 Feet High and Rising*, was a world away from the revolutionary screeds of Public Enemy, N.W.A., Ice-T, and Boogie Down Productions. It wasn't that they lacked for braggadocio, but these "phrasing Fred Astaires" brought sly finesse, a looseness, a tinge of jazz, and a suburban worldview that had not been seen before in hip-hop.

A SONG TITLE	AN ALBUM	A BAND	A LYRIC
"Three of a Perfect Pair" – King Crimson (1984)	*3-Way Tie For Last* – The Minutemen (1985)	Three Dog Night – L.A., pop-rock (1960s–2010s)	"We have many things in common, name three" – R.E.M., "These Days" (1986)

What's your favorite #4 song? Chances are, unless you are more of a numero-musical obsessive than I am, nothing leaps to mind. Most pop songs are written in 4/4 time, but #4 songs lack the semantic immediacy of "One is the loneliest number" or "Three is the magic number"—or even "When two tribes go to war." Fact is, "Four of Us" just doesn't have the same ring as "Two of Us" (although the Four Tops and John Sebastian went for it anyway).

Four has its associations, of course, even some signature acts to call its own (the 4-minute mile, breaking the fourth wall). But the essence of 4-ness is harder to pin down than the essences of the preceding trio. It has at least one unique property though: 4 is the only numeral with the same number of letters in its name as its value, a phenomenon that holds true across several languages. But compared with 1, 2, and 3, 4 is much more of a character actor than a leading man. The pop charts are littered with one-, two-, and three-titled hits, while only a smattering of four-titled songs have made it into the Top 40, the most recent being the egregiously catchy "4 Minutes" by Madonna and Justin Timberlake.

For most of the past century, "I'm Looking Over a Four-Leaf Clover" was the ultimate #4 song. Al Jolson sang it back in the '20s, and in 1948 it became a national hit for bandleader Russ Morgan. (I dig the warbled rendition by jazz guitar master Nick Lucas.) But that's your grandfather's four-leaf clover. "I'll give you a four-leaf clover/take all the worry out of your mind," sang Pete Townshend on "Let My Love Open the Door." And a healthy crop of four-leaf-clover songs has sprung up since then, from Old '97s, Badly Drawn Boy, Erykah Badu, Abra Moore, and Winger, so clover is by no means over. Metallica, however, issued a terse smackdown of the whole genre, "No Leaf Clover," replacing whimsy with the Book of Revelation, and their "Four Horsemen" uncannily suggests Sabbath's "Children of the Grave" played at 45 rpm. Judas Priest's "Four Horsemen," on the other hand, is oddly sedate. Most up my street is the Clash's "Four Horsemen." Feeling a bit doom-laden? Consider making a Four Horsemen of the Apocalypse mixtape, starting with the above and moving on to same-named songs by the Stranglers, Ralph Stanley, Glen Campbell, Aphrodite's Child, and the Klaxons. You could call it Four Horsemanure.

My first cursory mind-search for #4 songs yielded an appealingly random selection: Aphex Twin's gorgeously skittering "4"; "Four Sticks" by Led Zeppelin, perhaps the weakest link on *Zeppelin IV* but still pretty amazing; "Radio 4" by PIL, the stately, uncharacteristically restrained piece that closes *Metal Box* and wouldn't sound out of place nestled toward the end of Bowie's *Low*; and "December 4th" from Danger Mouse's *The Grey Album*, leveraging "Mother Nature's Son" in service of a moving narrative. But all of these seem to lack anything essentially 4-ish.

A couple of Italians named Tony and Frankie (Vivaldi and Valli) are more synonymous with the four seasons than any "Four Seasons"-named song, whether by Crowded House, Violent Femmes, Toots & the Maytals, Ambrosia or, for that matter, Killer Dwarfs or the Sadistic Mika Band. Four is the number of the bodily humors (blood, black bile, yellow bile, phlegm); the cardinal points (north, south, east, west); and the Noble Truths of Buddhism (they all involve suffering). But where are the songs to show for it?

The scads of "four walls" titles in existence testify to a more basic truth: that songwriters write songs in, and are oppressed by, four-walled rooms. Stevie Wonder's hard-luck protagonist in "Living For the City" grows up "surrounded by four walls that ain't so pretty." In Wings' "Band on the Run," Paul is stuck inside these four walls, sent inside forever. "Four Walls," by country crooner Jim Reeves, might be the darkest item in this claustrophobic subgenre. In a comforting baritone he sings that the walls are closing in, and you wonder why David Lynch hasn't used this track yet. In the world of jazz, 4 is often invoked to signify the 4/4 time signature or a variation on it. That was the reasoning behind the titles of two classics, Miles Davis's "Four" and "Four on Six" by Wes Montgomery (covered scintillatingly by pianist Wynton Kelly, a cohort of Davis).

4 FACT:
Beyoncé, Blues Traveler, Foreigner, Bloc Party, and Tupelo Chain Sex have little in common musically, but all have named records after the natural number following 3 and preceding 5.

The blues is well stocked with #4 songs. In "Four Day Creep," Ida Cox declares, "I'm gonna buy me a bulldog to watch my man while he sleeps." Robert Johnson's "From Four Until Late" details one of the choice segments of your typical four-day creep. "Tom Traubert's Blues (Four Sheets to the Wind in Copenhagen)" is a Tom Waits weeper with a "waltzing Matilda" refrain that tugs at the heart like a rainy day. Waits summed up its origins this way: "Uh, well I met this girl named Matilda. And uh, I had a little too much to drink that night. This is about throwing up in a foreign country." Canada, which many people consider a foreign country, checks into a double at the Motel 4, with BTO's "Four Wheel Drive" and Ian & Sylvia's "Four Strong Winds." The term "four-letter word" has officially signified a distinct class of taboo expressions since 1897. Simon & Garfunkel ruminated on "a single-worded poem comprised of four letters," aka the F-word, in "A Poem On the Underground Wall," but most songwriters take the position that the only four-letter word worth identifying as such is "love."[1] Joan Baez popularized Dylan's "Love Is Just a Four-Letter Word," whose title sentiment has been echoed by Cheap Trick, Def Leppard, Bon Jovi, and ace film scorer Roy Budd.

Much speculation has been devoted to determining whether the location in the title of Bob Dylan's "Positively 4th Street" refers to the Greenwich Village street where he once lived or an address in Minnesota. Almost as much has gone into identifying the target of the song's unremitting scorn. Some theorists pointed to Edie Sedgwick, while others suggested Dylan was railing against folk purists who were offended by his change of musical direction. Dylan's "Fourth Time Around," a song from *Blonde on Blonde* that's generally accepted to be something of a "Norwegian Wood" parody/homage, is just as inscrutable on the issue of 4-ness.

Songs that refer to the Fourth of July, or its more generic synonym, Independence Day, fall into two distinct camps of subject matter and spirit. Most Fourth of July songs deal with the long, lazy weekend that has come to symbolize the very heart of American summer. These are perfumed with grill smoke and suntan lotion, warm beer and fireworks. Songs that refer to Independence Day, however, are primarily concerned with personal, not national, liberation. The Beach Boys would seem tailor made for a Fourth of July song, and they do have one —but it's not the one you expect. Drummer Dennis Wilson, the Beach Boys' handsomest member as well as their only real surfer, wrote the stark, hymnlike "4th of July" to lambaste President Nixon. Accounts vary; either some of

---

1 The exception being Cilla Black's "Work Is a Four-Letter Word" (from a 1968 film of the same name), which was covered half-heartedly by the Smiths.

Wilson's fellow band members nixed the song from 1971's *Surf's Up* due to its dark message, or he withdrew it himself. In any case, by the time "4th of July" became generally available on a 1993 box set, Dennis Wilson had long since set sail.

Your ultimate "Fourth of July" song tells more about you than your favorite Christmas song ever will. So how do you roll? "Fourth of July" by U2? Come on, even rabid U2 fans are lukewarm about that one. Soundgarden's "Fourth of July," from *Superunknown*? Not really one of their best. Maybe you prefer the sepia-toned psychedelia of "Mister Fourth of July" by Joe Byrd and the Field Hippies. But I doubt it. By my calculations, you're either a cynic or a romantic. If you're the former, you're choosing "Fourth of July" by Galaxie 500, Dean Wareham's solitary bed-in commemoration of the national holiday, or Aimee Mann's "4th of July," in which she calls the whole thing "a waste of gunpowder and sky." If you're in the romantic camp, it comes down to two choices.

Dave Alvin's "Fourth of July" is a stirring slice of Americana that makes brilliant use of the title phrase, building to a rapturous chorus that rises like a cool wave through the heat. With just a few details, the song vividly conjures an unhappy couple for whom the holiday offers the possibility of redemption. In the version by X, Alvin's one-time bandmates kick the song up a few notches until it chugs like a classic rock anthem—I love them both.

THE VERDICT

"4th of July, Asbury Park (Sandy)" reminds us of how great Bruce Springsteen was, even before his legend was made with *Born to Run*. You can practically smell the salt air of New Jersey's cut-rate casbah, from the very first guitar frills all the way to the stirring climax, on which Springsteen on recorder uncannily conjures a chorus of seagulls, and a final phrase that almost wants to segue into the Crystals' "And Then He Kissed Me." From the Tilt-a-Whirl and the pier lights, to the boys from the casino dancing with their shirts open, the details give off the unmistakable waft of summer swoon. The song's sweetest image ("the aurora is risin' behind us") is one of the most poetic in all of Boss-dom, imbuing the heat-streaked essence of the Fourth with an aura of optimism.

A SONG TITLE	AN ALBUM	A BAND	A LYRIC
"Four or Five Times" – Sister Rosetta Tharpe (1941)	*Vol. 4* – Black Sabbath (1972)	The Four Tops – Detroit, R&B/soul (1950s–2010s)	"We've done four already but now we're steady, and then they went 1, 2, 3, 4!" – Led Zeppelin "The Ocean," (1973)

5

Our lives are full of fives. We humans have five sense organs and five fingers on each hand (the better to avail ourselves of a fifth of booze via the five-finger discount). There are five natural elements (air, earth, fire, water, and ether); five books of Moses, and Muslims face Mecca to pray five times a day.

Mostly we use 5 to refer to the slippery amount of time known as five minutes, which has given rise to an entire subgenre of songs about that elusive quantity: "Five Minutes of Fame," "Five Minutes of Flow," Five Minutes of Rage," "Five Minutes of Funk," "Five Minutes of Skunk," ad nauseam. In Sammy Cahn's chestnut "Five Minutes More," the singer begs his honey to extend their embrace by an additional one-twelfth of an hour; in Pantera's "Five Minutes Alone," the singer desires that same time period to pummel the crap out of his oppressors. But it's hard to top "Five Minutes" by Bonzo Goes to Washington, a collaboration between Jerry Harrison of Talking Heads and funk master Bootsy Collins, built around Ronald Reagan's famous 1984 ad-lib into an open microphone: "I'm pleased to tell you today that I have signed legislation that will outlaw Russia forever. We begin bombing in five minutes."

The Gipper, whose controversial quip came not long after his infamous "Evil Empire" speech, was roundly lambasted for amping up already-high tensions between the world's two superpowers. Enter Messrs. Collins and Harrison, who sampled and looped his ill-chosen words and placed them over some fresh beats. When the major labels balked, the pair successfully released it themselves, giving many people their first taste of sampling. The prescient Bootsy and Jerry could easily claim "I'm Five Years Ahead of My Time," after a 1967 track by trippily named The Third Bardo.

In "Five Days, Five Days," Gene Vincent laments the length of time since his baby walked out the door; in "5 months, 2 Weeks, 2 Days," Louis Prima does something similar, only Vincent sounds far more miserable than the exuberant Prima—which is odd, given that, relatively speaking, it should be the other way around. "5 Years" is a typically otherworldly concoction by Bjork, from *Homogenic*. She recorded it in the wake of her stalker's suicide and hoped it would sound like "rough volcanoes with soft moss growing all over it." Clearly, she succeeded.

"Nine Lives to Rigel Five" by the criminally unheralded '80s band Game Theory, refers to a star in the constellation Orion. "I checked the distance to Rigel," Scott Miller wrote in the liner notes for *Tinker to Evers to Chance*, an ostensible greatest-hits collection for his band, which never had any. "And it turned out to be very close to nine human lifetimes if you go at the speed of light." The song begins and ends with a backward recording of a vacuum cleaner, perhaps an unconscious nod to Joe Meek's homemade effects on "Telstar," the international hit from late 1962 that heralded the space age.

"Five Years," the stately lead track from *Ziggy Stardust and the Spiders From Mars*, is somehow easy to overlook on a record whose most recognizable highpoints are rockers fueled by Mick Ronson's pealing riffs.[1] Bowie channels an extraterrestrial rock star (the first of his many alter egos) and, through the cracked and compelling voice of Ziggy, conjures a dystopian world on the cusp of extinction. His exceptional vocal (which apparently took just two takes) starts off weary and resigned but finishes with anguish and urgency. Bowie has said that the idea for "Five Years" came to him in a dream in which his deceased father warned him that he had five years to live, and to avoid airplanes.

PLUS 5:
Proponents of the negativity bias theory tell us that in the average marital relationship it takes five compliments to make up for a single cutting remark. Something to keep in mind the next time you feel tempted to be brutally honest with your mate.

Five o'clock has become synonymous with the end of the working day, and many songs reference this association, from the big band hit "Five O'Clock Whistle" to the Jam's "Just Who Is the 5 O'Clock Hero?" and Jimmy Buffet and Alan Jackson's invitation to start drinking, "It's Five O'Clock Somewhere." The best of these is doubtless "5 O'Clock World" by the Vogues, a perfect slice of radio pop circa 1965, notable for its deeply catchy army-drill cadence and nifty yodeling bit in the turnaround. I actually discovered this song through Julian Cope's cover version, so listening to the original, without Cope's added lines about nuclear apocalypse and his interpolation of Petula Clark's "I Know a Place," was almost revelatory, like hearing "Hey Bulldog" in stereo for the first time, or noticing that the lady you brought home has 5-o'clock shadow.

MINUS 5:
…the Jackson …the Dave Clark …the Ben Folds …the Count …the Maroon …the Gramercy… the Crypt-Kicker[2]

Five bucks doesn't buy what it once did. In the jaunty George Jones-Gene Pitney collaboration "I've Got Five Dollars and It's Saturday Night," our boys have big plans for their fiver. "I'll give it five" used to mean something, too. Teenager Janice Nicholls sang a song named after this phrase, which she uttered to wide acclaim on the '60s British TV show *Thank Your Lucky Stars*. Ms. Nicholls, who later became a chiropodist, name-checks all the hottest acts of the day, including Chubby Checker, Alma Cogan, and Bobby Vee, in a strong Midlands accent that charmingly renders the title "Oi'll give it foive."

Can I get a high-five? Hey, don't leave me hanging. National High-Five Day (which celebrants commemorate by high-fiving everyone they meet) occurs on the third Thursday in April. Since 2002, the inventors of this holiday have also posted a list of suggested songs that lend themselves to the slapping of palms, including Bob Dylan's "Obviously Five Believers." But mystifyingly they left off a prime candidate in Beck's "High-Five (Rockin' the Catskills)." Are they kidding? Let me hear you say Sergio Valente!

I should probably take the advice of Dave Brubeck about now. His "Take 5" was the first jazz single to sell a million copies, and even non-jazz lovers recognize it when they hear it at the dentist's office. I wish I could say the same for "Take 5" by Northside, also-rans from the early-'90s acid house scene. They produced several cool singles that, it's safe to say, have never been played in a health care setting.

Sometimes a number is just a number. In songs like "5 from 13" by Soft Machine and "Five Per Cent For Nothing" by Yes, 5 is simply a mathematical value. In the Doors' "Five to One," sung by an audibly intoxicated Jim Morrison, the meaning of five

---

1 On his next album, *Aladdin Sane*, Bowie would revisit the song's chord sequence and 3/4 meter on the doo-wop flavored "Drive-In Saturday."
2 As in the Crypt Kicker 5, the vocal group of the coffin bangers in Bobby "Boris" Pickett's "The Monster Mash."

is a matter of conjecture. Some say the title refers to the ratio of young people to adults in 1967 America, or that of pot smokers to non-pot smokers, or Viet Cong to American troops in Indochina at the time. Whatever it means, this call to arms has been extremely influential: Jay-Z and Mos Def have both sampled it; Mike McCready of Pearl Jam based his guitar solo in "Alive" on Ace Frehley's solo in "She," which Ace "Frehley" admits nicking from Robbie Krieger's "Five to One" solo. And Oasis five-finger-discounted the tune on "Waiting For the Rapture."

The great blues shouter Jimmy Rushing was known as Mr. Five By Five, after a song he popularized about a guy who was "five feet tall and five feet wide." Elvis stood about six feet tall, and Johnny Cash, whose "Five Feet and Rising" recounted an actual flood from his childhood, was about 6'2". But many celebrated frontmen (Bowie, Bono, the Boss, Prince) are closer height-wise to Rushing than Presley. One of their great gifts is in seeming much taller than they really are.

Women of diminutive stature have more latitude. In "Five Feet of Lovin'," Gene Vincent raves that his five-foot-tall mama "is cool cool cool," and Gloria, muse to Van Morrison and Patti Smith, is "about five feet four/from her head to the ground." For most men, though, diminutive stature is rarely an asset, unless you make it one, like Johnny Rotten, who emanated menace in his debauched king's crouch, or Pablo Picasso, who was only 5-foot-3, but girls could not resist his stare.

THE VERDICT

Only Iggy Pop could sing "Five Foot One" from the point of view of a lovesick Lilliputian and get away with it. In the hands of anyone else, the song would come off as a joke, but Iggy turns this tale of an amusement park worker who longs to "go home with all the big folks" into the defiant cry of a wounded misfit. "I wish life could be Swedish magazines/I wish life could be … *anything*!" he howls before the song's chaotic fadeout.

"Five Foot One" appeared on Iggy's return to relative sanity, *New Values*, after several years of post-Stooges physical and emotional turmoil, and the urgency of his short-man protagonist reflects Iggy's newfound sense of purpose. Obviously, #5 is a crowded category, but the primitive power, snarling self-affirmation, and utterly singular worldview of "Five Foot One" make it my top choice, narrowly edging out Mr. Pop's friend and colleague Mr. Bowie. I mean, what the hell, what the heck??

A SONG TITLE	AN ALBUM	A BAND	A LYRIC
"Cubs in Five" – The Mountain Goats (1995)	*Five Leaves Left* – Nick Drake (1969)	The Jackson 5 – Gary, Ind., R&B/soul (1960s–1980s)	"Jenny said, when she was just five years old/ You know there's nothing happening at all" – The Velvet Underground, "Rock & Roll" (1970)

Six, like all its single-digit brethren, is everywhere, from country (Charlie Pride, "Six Days on the Road") and alt-country (Lucinda Williams, "Six Blocks Away") to Irish folk-punk (The Pogues, "Six to Go") and heavy metal (Alice Cooper, "Six Hours"). Ice-T's "6 in the Mornin'" has been called the first gangsta-rap song, and Sonic Youth weigh in no less than three times on six, most rockingly with "Six Hits of Sunshine (For Allen Ginsberg)." It's worth noting that most of the foregoing titles employ 6 as a measure of time or distance, not for any particular 6-ish properties. But a closer look at 6 reveals a number with a dangerous past.

Six has powerful associations with some very dark things, most notably weaponry (Nick Cave's "Six Inch Golden Blade," Queens of the Stone Age's "Six Shooter"), death (Hank Williams's "Six More Miles (To the Graveyard)"), and the classic Number of the Beast/Antichrist association ("666" by Anvil). But as long as guitars remain rock's essential piece of equipment, 6 will continue to be associated not with the great unknown but with rocking. Whether it's Bryan Adams's first real six-string, the loaded six-string on Jon Bon Jovi's back, or Mott the Hoople's six-string razor, songwriters continue to reference and romanticize the sunburst/candy apple tools of their trade.

In the 1970s came dire warnings that synthesizers were about to supplant guitars,[1] but guitars, and their six strings, have endured. Tributes to the instrument itself (Hank Snow's "Six String Tennessee Flat Top," Son Volt's "6 String Belief") are plentiful. Stevie Ray Vaughan's "Six Strings Down" is a tribute to "blues stringers" who have joined the choir eternal. Waylon Jennings writes about how a person's very survival can come to depend on those coiled wires in "Six Strings Away," and in the Birthday Party's "The Six Strings That Drew Blood," a young Nick Cave doomily intones the harrowing tale of "a guitar thug … forever the master and the slave of his six strings."

In volleyball, a six-pack is a spiked ball that slams an opponent in the face. Joe Six-Pack, formerly known as John Q. Public, is an average Joe who doesn't have six-pack abs. A six-pack may just be a delivery system for beer, but like the 40-oz., it has transcended that status and become a thing unto itself. The Jayhawks' "Six Pack on the Dashboard" calls up a classically American image, doesn't it? Six-packs have been saluted by both country music star Hank Thompson, in "Six Pack to Go," and Black Flag, whose scarifying "Six Pack" opens with "I've got a six pack and nothing to do." There's no six-pack in sight in "Six Feet of Snow," which finds one of Little Feat's archetypal heartsick truckers making the long journey home. Good thing too: "It's raining in stilettos from here clear down to Mexico."

---

1 The final credit listed on the back of Queen's *Sheer Heart Attack* (1974), for example, right below "sleeve concept by Queen," is "no synthesizers." In subsequent decades musicians have continued to foreswear various then-new means of making music. On the back cover of the Damned's damnably good 1982 LP *Strawberries* was the notation, "This is a synth-free album." Some kind of peak was reached with the egregious back-cover copy of Shriekback's *Big Night Music* (1986), in which the Shrieks declare: "It's perhaps worth mentioning that Big Night Music is entirely free of drum machines, sequencers, Fairlight Page R's – digital heartbeats of every kind. Seductive though they are, SHRIEKBACK [note dangling modifier] have opted to make a different kind of music – one which exalts human frailty and the harmonious mess of nature over the simplistic reductions of our crude computers." And as recently as April 2014, Win Butler of Arcade Fire told his audience at Coachella Music & Arts Festival to give a "shout-out to all the bands still playing actual instruments at this festival."

If 6 Was 9

For the sake of thematic consistency, it seems apt to distill the remaining offerings down to a six-pack of pure excellence. First up: Bob Dylan's "From a Buick 6," a screaming blues rave-up from *Highway 61 Revisited* that's cut from the same cloth as "Maggie's Farm." In this tribute to a "soulful mama" he extols a woman's charms (or possibly damns her with faint praise) as only Dylan can: "Well, she don't make me nervous, she don't talk too much/She walks like Bo Diddley and she don't need no crutch." Yo La Tengo tweaked Dylan's title in "From a Motel 6," one of the most heavy, and heavenly, numbers in their songbook. It's just huge-sounding, with the intertwined vocals of Ira and Georgia set against a thrumming wall of guitar noise. You can almost hear wings melting. Neither song is particularly interested in 6 though.

"Midnight to Six Man" (1965) is a rollicking celebration of late-night hedonism by the Pretty Things, rougher, tougher contemporaries of the early Stones. Like the Stones (and many bands of the era), they began the '60s as an R&B outfit, morphed into rock, and dived into psychedelia at the end of the decade. Debate still rages in some quarters as to whether or not the Who ripped off the Pretties' 1968 concept album, *S.F Sorrow*, for *Tommy*, which appeared a year later. The record has earned significant late acclaim, but the Pretty Things were doomed to also-ran status in their lifetime. Even with their 1970 album *Parachute* earning top honors in *Rolling Stone*, the band never had an American hit. Most stateside listeners first encountered the band's music when Bowie covered "Don't Bring Me Down" and "Rosalyn" on *Pinups*, but then, the Pretty Things were always a band's band. Steven Tyler and Noel Gallagher have cited them as key early influences, and not for nothing does a top-level Clash song begin with the words "Midnight to six man."

The Lovin' Spoonful are justly known for a run of great singles in the mid-'60s, including "Do You Believe in Magic," "Summer in the City," and "Daydream." The less familiar "Six O'Clock" is a gem from that same golden era, when cultural changes, abetted by advances in recording techniques, enabled the creation of miniature worlds in three minutes or less. Here, a staccato keyboard line that strongly echoes the opening of the Beatles' "Getting Better" is followed by John Sebastian's bell-bright tenor, bringing the dawn with the lines, "There's something special 'bout six o'clock/in the morning when it's still too early to knock." (For what it's worth, Paul McCartney eventually wrote his own "Six O'Clock," with a little help from his friend Linda, and gave the song to Ringo.) The opening keyboard figure of Scritti Politti's "After Six" has a caustic texture not unlike the keys in "Six O'Clock," but the similarities end once a galloping shuffle beat kicks in. It's doubtful that anyone will ever write a catchier ditty about rejecting Christianity. Scritti mastermind Green Gartside, it should be noted, stands 6-foot-6.

Measuring five inches shorter than Mr. Gartside is "6' 1'," the lead track from Liz Phair's crucial debut, *Exile in Guyville*.[2] Any great record needs a great beginning, and "6' 1'"—a gimlet-eyed evisceration of a man who beds girls who are "shyly brave"—sets up *Guyville* beautifully. Liz Phair's sly, sexually frank lyrics initially stopped listeners in their tracks, but the talk got less interesting with time. Singing "I want to be your blowjob queen" was audacious in 1993; 10 years later, naming a song "H.W.C." (as in "hot white cum") felt decidedly less so. I could never actually hear how the record correlated musically with the Stones' *Exile on Main Street*, but one thing's for sure: the rock 'n' roll boys club was never the same again.

---

2 Liz released another #6 song, "Six Dick Pimp," on one of the cassettes she put out under the name Girly-Sound.

David Klein

THE VERDICT

Fine offerings all, but "If 6 Was 9" by Jimi Hendrix, a marvel of controlled chaos and innovation by one of the giant figures in rock, has to take the prize. Prominently featured in *Easy Rider*, "If 6 Was 9" sums up the countercultural spirit of rebellion far more succinctly than the film itself. Indeed, as a signifier of the psychedelic pswirl of 1967, the song is unparalleled. The makers of *Mad Men* used it to signify personal liberation for an episode in which the embattled Don Draper embraces his destiny and returns to the fold after a lengthy exile. The song survived a frame-by-frame remake by Todd Rundgren with its mystery intact and introduced the concept of the freak flag ("I'm gonna wave my freak flag high"). Besides the mind-blowing sonics, it also features one of Hendrix's most impassioned vocals—an aspect of his legacy that's usually given short shrift. (Hendrix himself was never comfortable with his singing voice.)

In a 2008 NPR broadcast, Adrian Utley of Portishead talked about the impact of hearing "If 6 Was 9" in 1970 as a budding beatnik of 13. "The sound was so vicious and brilliant," said Utley, who was especially taken with the line, "If all the hippies cut off all their hair/I don't care"—a surprisingly contemptuous sentiment at the height of flower power. From a numerological standpoint, "If six turns out to be nine, I don't mind" is a powerful, even deep incantation, a questioning of the truths we hold dear, an affirmation of selfhood, a riff against complacency.[3] Hendrix looks at the number 6 and sees the Great What-if, befitting a man who similarly reimagined conventional notions of guitar playing, and whose accomplishments on the Stratocaster remain the absolute pinnacle,[4] like the first man on the moon.

Wave on, wave on …

A SONG TITLE	AN ALBUM	A BAND	A LYRIC
"Six Day War" – Colonel Bagshot (1971)	*Beatles VI* – The Beatles (1965)	Six Organs of Admittance – California, indie folk (1990s–2010s)	"Robbing people with a six-gun/ I fought the law and the law won" – The Bobby Fuller Four, "I Fought the Law" (1965)

---

3 Compared with Z.Z. Top's "I got the six, gimme your nine," it's Shakespeare, but They Might Be Giants tweaked Hendrix's postulation rather adroitly in "Secrets of Six," imagining 6 "when he pretends to be 9." Hendrix acolyte Rick James offered up his own twist in "You and I," declaring, "We'll be together 'til the six is nine."
4 Not everyone would agree. Robert Fripp said: "You know Hendrix? He didn't know how to hold a pick."

Is 7 a blessed number that brings good fortune? Well, if you ask the 60,000 couples worldwide who chose 7/7/07 as their wedding day, the answer, after "I do," would be a resounding affirmative. The love of 7 is a universal phenomenon and has been for ages. "Seven is the number of the young light," reads the I Ching, "and it arises when six, the number of the great darkness, is increased by one."[1] Seven is associated with perfection and completeness in all major religions (see the Old Testament, the Kabbalah, the Pixies' "Monkey Gone to Heaven," and other holy texts). It even holds true for not-so-major religions (Zoroastrianism, anyone?). Then again, according to the New Testament, there are seven signs of the apocalypse, so go figure.

The seven deadly sins—lust, gluttony, greed, sloth, wrath, envy, and pride—have been around in various forms since the fourth century and have made their way into verses by Dante and Chaucer, paintings by the likes of Hieronymus Bosch, a "sung ballet" by Kurt Weill and Bertolt Brecht, and a concept album by Joe Jackson. The seven cardinal virtues—faith, hope, charity, etc.—have inspired nothing approaching the creative outpouring wrought by the sins (although "Charity, Chastity, Prudence and Hope" by Hüsker Dü is pretty cool). Nothing against Flogging Molly, Gene Loves Jezebel, the Traveling Wilburys, Simple Minds, the Dubliners, and the many others who've committed "Seven Deadly Sins" to vinyl (or another recording medium), but I'm keenest on the debut single by Brian Eno, freshly jettisoned from Roxy Music in 1974.

OK, "Seven Deadly Finns" is just a punning reference to the sins, but that's close enough for me. In this blissful song about Finnish sailors thrilling bored French women, Eno gives each sailor a specific attribute: There's the masochistic freak, the treed kitten, the outgoing cross-dresser, the Eno impersonator, the distrustful hat enthusiast, the indoors sunglasses type, and the skinny outcast. The giddy enthusiasm of the song is so joyous, the only logical conclusion is to erupt into yodels. Indeed, no finer instance of yodeling in a rock song exists (except possibly "Hocus Pocus" by Focus).

Seven-related phenomena come so thick and fast that one reference builds upon another. "The Magnificent Seven," the Clash's first foray into rap, takes its name from the classic 1960 western, which was modeled on Kurosawa's *Seven Samurai*. "The Seventh Seal" is Scott Walker's musical retelling of the 1957 Bergman film of the same name, whose title comes from a passage in the Book of Revelation: "And when the Lamb had opened the seventh seal, there was silence in heaven about the space of half an hour." Clocking in at well under the space of half an hour is the cult classic "7 Screaming Dizbusters," if by cult you mean Blue Öyster with an umlaut.

According to legend, the seventh son of a seventh son is destined for greatness. Somewhat ironically, human sleeping pill Perry Como was a real-life seventh son of a seventh son. But references to 7 are a staple of the blues, with Muddy Waters' "Hoochie Coochie

---

1   Pink Floyd adapted these and other lines from the I Ching in "Chapter 24" on *The Piper at the Gates of Dawn*.

Man" ("On the seventh hour/on the seventh day/on the seventh month/the seven doctors say") and Willie Dixon's "The Seventh Son" representing the ultimate seven-to-the-seventh-power tracks. "Seventh Son" is one of music's great boasts. Not only is the title son a lover beyond compare, he can also heal the sick and raise the dead. The protagonist in Iron Maiden's "Seventh Son of a Seventh Son" is also a healer type, but his sexual prowess goes unmentioned. A line in Dylan's "Highway 61 Revisited" ("But the second mother was with the seventh son") is believed by some Dylanologists to be an incest reference, but to paraphrase Bill Clinton, it all depends on how one defines "with."

An aspect of 7 that has been manna to songwriters over the years is that it rhymes with heaven. In Islam, the heavens number seven. While it's unclear if the term "seventh heaven" has an Islamic origin, the association is drilled into us from childhood sing-alongs of that old counting chestnut "This Old Man," and it has provided song titles for heavy acts like Deep Purple and Prodigy, as well as Gwen Guthrie of "Aint Nothin' Goin on But the Rent" fame. More generally, it's tough to find instances of 7 that *aren't* rhymed with "heaven." Songs by Bill Haley & His Comets ("When the chimes ring five, six, and seven/we'll be right in seventh heaven"); The Beatles ("1, 2, 3, 4, 5, 6, 7/ All good children go to heaven"); and the Ramones (Four-five-six-seven/ All good cretins go to heaven!") have earned the seven-heaven rhyme prominent placement in the Big Book of Pop Lyrics.

Seven-11, a winning combination in dice games, appears frequently in blues and cowboy songs, while the Ramones made good use of the term's convenience-store connotation in a song whose refrain goes "I met her at the 7-Eleven/ Now I'm in seventh heaven." Undoubtedly the worst of the many seven-heaven songs is Paul Nicholas's über-melted-cheesy "Heaven on the Seventh Floor," a 1977 hit for a performer who started his career in 1960 as Paul Dean, then went simply by "Oscar" before coming up with a lasting stage name. He played Jesus (as in *Christ Superstar*) on the London stage, Cousin Kevin in Ken Russell's *Tommy*, and also served as TV pitchman for the dubious Rougemont Castle wine, which is, to borrow a phrase from Monty Python, "an appellation contrôlée specially grown for those keen on regurgitation."

A number as ubiquitous as 7 yields a bevy of tracks called simply "Seven" (or "7"), and an accomplished bevy it is. There's "Seven" by David Bowie, "Seven" by Dave Matthews Band, and "7" by Prince, a gospel-tinged sing-along with lapidary production touches that imagines the eventual demise of the seven deadly sins. A more recent vintage is "Seven" by Fever Ray, the solo project of the Knife's Karin Dreijer Andersson, whose voice seems to emanate from an uncharted realm.

"And on the seventh day He rested," a Bible quote, is responsible for the seven-day week as well as a slew of related song titles.

In Bob Dylan's initially unreleased "Seven Days," the desperate singer awaits the arrival of a woman whose face could outshine the sun in the sky—all he has to do is survive. A similar sentiment of longing pervades The Four Tops' "Seven Lonely Nights," which finds Levi Stubbs crying, dying, and sighing all week for his baby, and Chuck Wood's "Seven Days Too Long," a Northern Soul barnburner covered by Dexy's Midnight Runners. The Dubliners' "Seven Drunken Nights" celebrates whiskey-soaked abandon, while "Seven Nights to Rock" is a thumping proto-rocker by Moon Mullican, who claimed he took up the piano "because the beer kept sliding off my fiddle."

More 7-ness? "Seven Yellow Gypsies" by British folk music patriarch Martin Carthy is a favorite of Robyn Hitchcock, who told me that in an email. "Seven Ways to Jack" is an early house single by Hercules. "(Seven Little Girls) Sitting in the Backseat" is a corny '50s novelty. There are highbrow offerings from the Teardrop Explodes ("Seven Views of Jerusalem"); Jane Siberry ("Seven Steps to the Wall"); and Aphrodite's Child ("Seven Trumpets"); and proggy things from Genesis ("Seven Stones") and Adrian Belew ("Seven E-Flat Elephants"). Liz Phair's "Dance of the Seven Veils" contains one of rock's finest instances of the C-word. "7 Seconds" was a European hit for Youssou N'Dour and Neneh Cherry (Ms. Cherry lost the 1990 best new artist Grammy to Milli Vanilli—oh, the irony). Sting's "Love Is the Seventh Wave" was an overly optimistic forecast from *Dream of the Blue Turtles*, while Smashing Pumpkins' "7 Shades of Black" finds Billy Corgan entreating someone to "fall in hate with me."

Truly a "Smoke on the Water" for the 21st century, the White Stripes' "Seven Nation Army" is so good it threatens to overshadow much of this beloved duo's recorded work. The title is based on a youthful mishearing of "Salvation Army" by John Anthony Gillis, before he adopted a tri-toned wardrobe and changed his name to Jack White. Indicative of its outsize stature, "Seven Nation Army" has spawned tributes: a dub version by Hard-Fi, an electro remix by JAS-3, and a monolithic workout by Metallica, all serving to highlight the allure and versatility of the song's central seven-beat phrase, not to mention White's gloriously paint-peeling vocals and wailing slide licks. You can imagine that in a hundred years this'll still sound like all hell breaking loose.

As a young boy, John Gillis might well have come across *The Five Chinese Brothers* by Claire Huchet Bishop, a top-selling children's book for many years but now out of favor due to its brazen stereotypes and grisly subject matter. It's the tale of a Chinese man wrongly accused of murder who escapes various methods of execution by having each of his identical, but uniquely talented brothers take his place in turn. I'm sure it haunted young Michael Stipe of R.E.M., whose "7 Chinese Brothers" (two siblings were added, presumably for reasons of cadence) is a reminder of just how distinct indistinctness can be. Stipe's vocals, the solid ensemble playing, and the slowly unfolding layers of Mitch Easter's detailed production all cast a powerful spell. All hail the autumnal glory of early R.E.M.

The seven seas have inspired a range of artistic expression, from worthy (a collection of poetry by Rudyard Kipling) to much less worthy (the last song on the last album by Flock of Seagulls). With apologies to OMD's "Sailing on the Seven Seas" and Queen's "Seven Seas of Rhye," give me "Seven Seas" by Echo & the Bunnymen. A standout from the group's most fully realized record, *Ocean Rain*, "Seven Seas" is pure seduction, a monument of sumptuous orchestral pop topped with a characteristically swaggering vocal by Ian McCulloch, who manages to turn kissing a tortoise into an act of transcendence. It's both grand and grandiose, the embodiment of how to get away with making an outsize gesture in the context of a rock record. When the Bunnymen toured in support of *Ocean Rain*, the sugar-lipped frontman was wont to introduce new selections with, "This is another new one … off the greatest album ever made."

David Klein

THE VERDICT

My allegiances have shifted spasmodically in an effort to bestow the #7 crown. The Four Tops have *three* #7 songs. Besides the aforementioned "Seven Lonely Nights," "Seven Rooms of Gloom" was a 1967 hit that's almost operatic in its dark drama. The less known "Just Seven Numbers (Can Straighten Out My Life)" also features a powerhouse vocal from Levi Stubbs, along with a celestial oboe solo and the sound of a rotary dial phone on the fadeout. All three of these heartbreakers are gorgeous and worthy of the crown.

Still, I'm currently inclined to award the prize on "7 and 7 Is," a 1966 single by Love and the band's only hit. It reached No. 33 on the *Billboard* chart in July of that supremely cool year. A two-minute sprint culminating in the sound of a nuclear explosion (replete with a countdown), "7 and 7 Is" has been rightly called proto-punk. With its unrelenting forward push, minor chords, and hostile protagonist, the song feels spiritually related to "Paint It Black," which had topped the charts in June.

Led by the Memphis-born Arthur Lee, Love briefly ruled the L.A. rock scene, only to be supplanted by the Doors, with whom Love shared a producer, an engineer, and a record label. While the Stones were expected to be provocative, Lee, with his band's hippie-pleasing moniker and his own penchant for writing sweet melodies, could easily have taken the path of least resistance and ridden the burgeoning psychedelic ethos of the day. But Lee's contrarian streak is well documented. This is, after all, the man who redid Bacharach-David's "My Little Red Book" in primitive garage rock style and turned down an invitation to play at the Monterey Pop Festival, purportedly because of his intense dislike for the show's producer, Lou Adler, but more likely, according to author Barney Hoskyns, because the band had descended into "a state of drug-crazed chaos." Peter Albin, guitarist for their contemporaries Big Brother and the Holding Company, said simply: "Love should have been called Hate."

Somehow the band still managed to produce its masterpiece, *Forever Changes*, several months after passing up that epic rock festival. If Lee had played his cards right, Love would be remembered as one of the great bands of the '60s, rather than a group cherished solely by rock's most discerning contingent of listeners. But popularity was never a major concern for Arthur Lee. In a different song, he summed up his stance: "Sometimes I deal with numbers/And if you wanna count me//Count me out."

A SONG TITLE	AN ALBUM	A BAND	A LYRIC
"#7" – Polyrock (1980)	*Seven & the Ragged Tiger* – Duran Duran (1983)	L7 – L.A., grunge (1990s)	"Then God is seven/ then God is seven" – Pixies, "Monkey Gone to Heaven" (1989)

POSTSCRIPT: The B-side of "7 and 7 Is," "Number 14," is a response to the sum inherent in the A-side's title. Rock scribe Chuck Eddy described it as "perhaps the only Band-style Civil War rebel-nostalgia ever sung by a descendant of slaves."

On 8/8/08, *The New York Times* ran a piece called "Crazy Eights." Readers learned about the "deranged" Roman emperor Elagabalus, who held octal-themed dinners to which he'd invite eight very tall men, eight men with gout, eight men with hooked noses, and so on. Mary Queen of Scots, we learned, decreed that no one with a rank lower than earl or archbishop could eat more than eight dishes at one meal. Rather than try to compete with such erudition, I just tip my eight-cornered hat, offer up a toast (V8, naturally), and proffer my own list of 8-themed musical associations.

"Eight Arms to Hold You" was the original title of the Beatles' *Help*. The 8-track, an endless loop of 1/4" magnetic recording tape, is a low-fidelity icon created by Bill Lear, who also invented a jet or two. The boogie-woogie bugle boy of Company B played his horn "eight to the bar." *Tobor the Eighth Man* was an American adaptation of *8-Man*, a Japanese cartoon from the mid-'60s starring what's considered the first robotic manga character. Tobor ("robot" spelled backward) derived extra strength from smoking "energy cigarettes." (Of course these days, any purveyor of children's entertainment who suggested such a plot point would be declared a Section 8—the Army term for a soldier who is too mentally addled to participate in war.)

In "I'm Henry VIII, I Am"—a rare Roman numeral entry—Herman's Hermits switch out the tyrannical Tudor for an ardent stepdad. Based on a hoary English music-hall ditty, this less-than-two-minute earworm sold in record numbers in 1964, but the Hermits didn't think much of it. The song helped break them in America but underscored their lightweight reputation, a source of tension throughout the band's brief but successful run.

Eightball, the brilliant '90s comic by Daniel Clowes, probably takes its title from the expression "behind the eight ball." A billiards term meaning in a disadvantageous position, the phrase has provided song titles for the likes of Bill Haley, Madness, and Lee Dorsey. The Jody Grind's "Eight-Ball" draws upon the phrase's billiards implications, while songs referring to one-eighth of an ounce of cocaine, such as NWA's "8 Ball" and Super Furry Animals' "Baby Ate My Eightball," could fill a large baggie. Less Than Jake's "Ask the Magic 8 Ball" refers to the Mattel toy that's been answering yes-or-no questions with oracular authority since the 1940s, but Underworld's "8-Ball," from the soundtrack to *The Beach* (2000), should be sought out at all costs, the one true diamond connected with that piece of cinematic hokum.

The protagonist in Alfred Bester's *The Demolished Man*, which won the inaugural Hugo Award in 1953, plots a murder in a future where telepaths monitor the minds of the populace to prevent crimes before they can occur. To keep from being "read," he hums this disturbingly catchy jingle:

> *Eight, sir; seven sir;*
> *Six, sir; five, sir;*
> *Four, sir; three, sir;*

David Klein

Two, sir; one!
Tenser, said the Tensor.
Tenser, said the Tensor.
Tension, apprehension
And dissension have begun.

The Liars incorporate the sequence in "The Pillars Were Hollow & Full of Candy So We Tore Them Down" (an #8 song only in the loosest sense, but too cool not to mention). The following, however, is a bona fide 8-pack of primo octo:

- ✷ "Eighth Avenue," Hospitality's dreamy, number-sprinkled evocation of a sublime Manhattan that already feels like a thing of the past
- ✷ "Eight Men, Four Women," in which Memphis soul man O.V. Wright begs the jury for a last-minute plea deal in the courtroom of the heart
- ✷ "8:05," a gauzy gem from the egregiously talented, short-lived Moby Grape
- ✷ Neu's "After Eight," like a New York Dolls stomper tossed headlong down the autobahn
- ✷ "Eight Piece Box" by Southern Culture on the Skids—the last word in woman-as-roadside-meal imagery
- ✷ "Figure Eight," a sensual math lesson sung by the wonderfully understated jazz chanteuse Blossom Dearie
- ✷ "Eight Line Poem," David Bowie's country-tinged surrealist folk song, which asks, "Will all the cacti find a home?"
- ✷ "Eight Days on the Road," a versatile nugget recorded by Howard Tate and covered by Aretha Franklin and Foghat.

Despite a pretty rich turnout, the #8 slot comes down to a cage match between two jangly staples of the soundtrack of the 1960s. The question is this: Are you a Beatles person or a Byrds person? In one corner, "Eight Days a Week"—Lennon and McCartney at the peak of their collaborative powers, voices melding in a delirious, rough-hewn harmony of rare beauty. Recorded in October 1964, the song represented a couple of firsts for the Beatles. They had never before taken an unfinished sketch of a song and experimented with various ways of recording it in the studio. Those swelling, heavenly chords that seem to arrive out of nowhere to mark the song's beginning—pop's first fade-in—came about during remixing sessions, well after the best take (no. 13). The end result of all that tinkering? A song that simply has everything: the ringing guitars that launched a thousand jangles, perfectly deployed handclaps, words that simply yet purely evoke youthful romantic infatuation, all packed into a bracing 2:44. And appropriately enough, the song has a world-class middle eight. Alma Cogan, the Runaways, and the Dandy Warhols all attempted "Eight Days a Week," to no avail. Although the Beatles themselves were not especially enamored of it, and never played it live, the song is untouchable.

But maybe the whole "Ooh, I need your love, babe" thing is no longer relevant to your nuanced existence. Perhaps you prefer the challengers, in the fringed corduroy trunks and rectangular violet shades. From the ominous distorted bass line that opens the Byrds' "Eight Miles High" like a Morse code signal, joined by those unmistakable harmony vocals and Roger McGuinn's signature 12-string Rickenbacker guitar sound—Coltrane lines squeezed through a lysergic Play-Doh factory—this is as close to capturing the drug experience on record as we get. While it spawned such formulaic psychedelia as the Lemon Pipers' "Green

Tambourine," this McGuinn-Gene Clark-David Crosby creation conveys both the euphoria and paranoia that an acid trip can bring. For such a singular concoction, it has lent itself surprisingly well to other treatments. Hüsker Dü's demolishment of "Eight Miles High" is one of the band's recorded highlights, and even Roxy Music's disco-ball take on *Flesh and Blood* weirdly works. Living up to its name, "Eight Miles High" is hard to top.

THE VERDICT

In a certain sense it's pointless to say one is better than the other, unless it makes sense to argue that a lemon is better than a lime, a crocus is better than a snapdragon. I'm tempted to consider these old warhorses akin to Dame Judi Dench and Helen Mirren canceling each other out at the Oscars. *And the award goes to … Anna Paquin* (in the form of R.E.M.'s "Driver 8"). Great song, of course, but how could R.E.M. win out over the Byrds, when R.E.M's trademark sound is, like, two-thirds Byrds? Yet the Byrds themselves took ample inspiration from the Beatles, including in the naming of "Eight Miles High," which started out as "Six Miles High" until Mr. McGuinn added two miles in honor of the Fab Four song. With my back to the wall, I have to give the nod to the Beatles, using the following criterion: If stuck on a desert island with a record player and a vinyl 45 of one of these two songs, I would opt for "Eight Days a Week." Its endless optimism might spur me to carry on, while the crazed climax of "Eight Miles High" would surely render me a Section 8.

A SONG TITLE	AN ALBUM	A BAND	A LYRIC
"Eight Pictures" – The Go-Betweens (1981)	*The Eight-Legged Groove Machine* – The Wonder Stuff (1988)	808 State – Manchester, England, techno (1988–2000s)	"Got sent up for a eight-year bid" – Grandmaster Flash, "The Message" (1982)

Nine is about good things: the players on a baseball team, Salinger's *Stories*, a cat's lives, the number of the muses, being dressed to the nth degree, the chorus of "London's Burning" by the Clash—as I said, good things. Over the years, the third square number has inspired writers to produce an abundance of nine-centric songs, spanning decades, genres, and styles, and many of them are inarguably great. Sure, there are clunkers, like "Nine Tonight," by Bob Seger at his most lunkheaded (and I *like* Bob Seger). And why are #9 songs about the classic office hours, like Dolly Parton's "9 to 5" and Sheena Easton's "Morning Train (9 to 5)," so dull and workmanlike? In "Lady Marmalade," Allen Toussaint makes no bones about positing the work hours as the polar opposite of gitchi gitchi ya ya ta ta ("Now he's at home doin' 9 to 5"/Back to his gray flannel life"). Even the Ramones picked up on this sad trend with "It's Not My Place (In the 9 to 5 World)," from one of their lightest albums, 1981's *Pleasant Dreams*. The Kinks' "Nine to Five" is no great shakes either.

In "Love Potion #9," popularized by the Clovers, a visit to a fortune-teller transforms "a flop with chicks" into a cop-kissing, romantic fool. It's prototypical Leiber and Stoller, who liked to marry a memorable melody to a story-song that spoke to the kids. In his 2009 memoir, *Hound Dog*, Mike Stoller cites their earlier "Riot in Cell Block #9" as an inspiration for "Love Potion," noting that he and his partner were fond of 9 "because it resonates in song."[1] Like other merry melodies that depict magical transformations wrought by the ingestion of mysterious elixirs, "Love Potion #9" was rumored to be a disguised drug song, a notion that Mr. Stoller emphatically rejects.

The traditional "Nine Pound Hammer" dates back to the 1920s and is credited to Merle Travis, who introduced it to the masses in 1947. Travis, a triple-threat singer-writer-innovative guitarist, turns in a marvelous version of the song on 1972's *Will the Circle Be Unbroken*, the 3-LP collection under the auspices of the Nitty Gritty Dirt Band that introduced the Carter Family, Doc Watson, and other country legends to mainstream audiences, me included. The last line of the song's final verse ("You can make my tombstone/out of number nine coal") refers to a specific Western Kentucky coal seam bearing that number. Travis returned to the subject in "Sixteen Tons," an even bigger hit.

"Revolution 9" is as hard to love as anything the Beatles ever released (with the possible exception of "Mr. Moonlight"). The drearily intoned title numeral, taken from tapes for the Royal Academy of Music that John Lennon dug up at Abbey Road Studios, sets the stage for an otherworldly sound collage that incorporates symphony recordings, a slice of orchestra from "A Day in the Life," a backward Mellotron, and random spoken words from found sources. As Ian MacDonald pointed out in the indispensable *Revolution in the Head*, "'Revolution 9' provided the pop-buying public with its first exposure to experimental

---

1 Leiber and Stoller liked to use 9 alliteratively, but it also lends itself to assonance, à la Jimmie Rodgers' "Blue Yodel No. 9" ("Nine mile skid on a 10 mile idle").

techniques developed by Stockhausen and John Cage and the cut-up texts of William Burroughs—and probably its last." MacDonald further postulated that "Revolution 9" is the only Beatles song that most people only listened to once. The resonant quality of "number nine" that Leiber and Stoller liked so much, and which Lennon fully exploited, for better or worse, on "Revolution 9," can also be heard on songs like "The Wreck of the Number 9," a chilling tale of death on the rails sung by Tex Ritter, Hank Snow, and Jim Reeves.

The number 9 figures prominently in the life of John Lennon. Among other 9-related phenomena, he was born on the ninth of October, as was his son Sean; the Beatles were offered their first recording contract, with Parlophone, May 9, 1962; he met Yoko Ono on November 9, 1966. Eight years later, Lennon had a big hit with "#9 Dream." According to May Pang (in *Loving John*, her memoir of their 18-month, Ono-sanctioned affair),[2] the song's strangely addictive chorus of "Ah! böwakawa poussé, poussé" came to him in a dream in which two women were calling his name and intoning this exotic sounding phrase. In the recording, Pang is the one whispering "John." One year later, Lennon collaborated with David Bowie, whose "Five Years" also came to him in a dream. Five years later, Lennon was gone.

R.E.M., who covered Lennon's "#9 Dream" for a benefit album in 2007, seem to share Lennon's fondness for the number 9. The raw punk energy of "9-9" is somewhat at odds with the rest of the classic *Murmur*. It has been argued that the overall quality of an R.E.M song is in inverse proportion to the number of decipherable lyrics. By that standard, "9-9," which opens with the purely textural lyric, "Steady repetition is a compulsion mutually reinforced," ranks right up with their best. When the band announced it was "calling it a day" after 31 years (on 9/21/11), many people responded by posting lists of their favorite R.E.M. songs. Several that I saw, rather unexpectedly, included "9-9."

Guns are primal iconography across musical genres and time periods, but 9mm handguns appear almost exclusively in the world of hip-hop. (Lou Reed's "The Gun," in which he gives a positively chilling enunciation of the words "nine millimeter Browning," is a rare exception.) The term crops up frequently in the 1980s work of Ice-T and Boogie Down Productions ("I got myself an Uzi and my brother a 9") and has shown no signs of slipping. On David Banner's "9mm," Lil Wayne raps, "I got a girl, you wanna meet her/Her name is nine millimeter."

The caliber is unspecified in Drive-By Truckers' "Nine Bullets," in which Patterson Hood declares in a scorched tenor, "My roommate's gun's got nine bullets, I'm gonna find a use for every last one." After the bridge, Hood calls out the bullet earmarked for his immediate family, his voice ascending to a tone of raw pleading. But the darkness is offset by an unexpected "shave and a haircut" riff at the song's tail end, suggesting that maybe, just maybe, he's only kidding.

Far from the Truckers' Southern climes lies "9th & Hennepin," a brief spoken-word piece from Tom Waits's *Rain Dogs* that finds him on his familiar piano bench in the saloon at the end of the world. The title address is an actual corner in Minneapolis, and most of the imagery is New York City-based, but the result is pure Waits. "Well it's 9th and Hennepin, all the donuts have names that sound like prostitutes," he rasps above discordant twinkles and the eerie sound of a bowed saw. Possibly around the corner, at least in spirit, is "Apartment No. 9," the sunless abode where Tammy Wynette sits and waits, no doubt in vain, for her man to return. Tammy had her first hit with this Johnny Paycheck weeper.

---

[2] Harry Nilsson, Lennon's compatriot during that infamous "lost weekend," chose "Nine" as the middle name of his first-born son, Zak. Other musical believers in numerical naming include André 3000 of Outkast and Erykah Badu, who named their son Seven.

The phrase "cloud nine" has been around since the mid-20th century. The first documented instance of the term might well be in a 1935 slang compendium called *The Underworld Speaks*, which included the expression "cloud eight," as in: "Cloud eight, befuddled on account of drinking too much liquor." In the next decade, clouds seven and thirty-nine appeared; indeed, it was "cloud seven" that first found its way into an official slang dictionary, in 1960. But one has to assume that by 1969, the start of master producer Norman Whitfield's "psychedelic soul" period, cloud nine had supplanted cloud seven, once and for all.

"Cloud Nine" marks a major shift for the Temptations, from the catchy, well-tailored Motown sound of the mid-'60s to something harder-edged and hallucinatory, closer in spirit to Sly & the Family Stone than the Four Tops. The guitars are wickedly fuzzed-out, and lyrically it's a world away from expressions of love and lust like "My Girl" and "Ain't Too Proud to Beg." Whitfield himself initially resisted the edgier aesthetic, but when he embraced it the result was this chart-topping, Grammy-winning nugget, which has to rank with the Temps' best. With barely contained urgency, Melvin Franklin, Dennis Edwards, Eddie Kendricks, Paul Williams, and Otis Williams trade vocals in a manner that would find its fullest expression a few years later in "Papa Was a Rolling Stone." While psychedelic soul's time has come and gone, this ode to the dark side of euphoria still packs quite a punch.

So does Wilson Pickett's take on Gamble and Huff's "Engine No. 9," which is funky beyond compare. Over an elemental groove—heavy on the cowbell, abetted by maraca, shakers, vibraslap, and the limber bass lines of Ron Baker—Pickett presides in his inimitable way, exhorting the players not to stop: "Oh, this soundin' all right. I think I wanna hold it a little bit longer." They do, for three more minutes marked by throat-shredding howls, cries of "Lord have mercy" and "Git it!" and "Play your guitar, son," and entreaties to "just keep moving." And Pickett makes full use of the alliterative and tonal qualities of that phrase—number nine—in a way that Mike Stoller would no doubt smile upon.

THE VERDICT

This is a tough call. Two slabs of sizzling soul: one a loose vamp, the other marked by temporal concision and a breadth of sonic touches. It's the choice between two exquisite ice creams—you wouldn't want to live in a world without either. If pressed, I'll give the nod to the Temptations, the wah-wah pedal prevailing over the cowbell, in a soul-shaking squeaker.

A SONG TITLE	AN ALBUM	A BAND	A LYRIC
"Drivin' on 9" – The Breeders (1993)	9 – Public Image Ltd. (1989)	Nine Inch Nails – Cleveland, Ohio, industrial dance (1989–2010s)	"Out of nine lives/ I spend seven" – The Band, "The Shape I'm In" (1970)

Not nine. Not eleven. Ten commandments. No wonder 10-named songs are a solid lot: they are linked inextricably to the very basis of Judeo-Christian morality. Granted, it's a heavy subject to tackle head-on in a pop song, but in "The Ten Commandments of Love," a valentine to fidelity and lasting romance, legendary doo-woppers the Moonglows stirringly suggest a concept that an average 1950s teenager could get cozy with. (True, the 'Glows only enumerate nine commandments of love, but the background vocals cunningly fool the ear into thinking it's heard the full decalogue.) In "10 Crack Commandments," Biggie Smalls offers ageless advice to the budding distributor of illicit substances. On a related note, 10-year jail sentences befall both the naïve country boy in Stevie Wonder's "Living For the City" and the lunatic in Warren Zevon's "Excitable Boy."

Harry Nilsson, no stranger to numerically titled songs ("Take 54," "1941," "One"), used the Ten Commandments as the basis for "Ten Little Indians." His source was the same short poem-turned-schoolyard jingle that Agatha Christie borrowed for the title of one of her most popular mysteries. (The original title, published in the U.K. in 1939, used an appalling racial epithet instead of "Indians," but the U.S. edition carried the title *And Then There Were None*.) Unlike the poem, in which each little Indian dies from a different random misadventure, in Nilsson's version[1] each one dies by breaking a commandment. The Beach Boys' "Ten Little Indians," one of their least successful singles, and deservedly so, uses the traditional sing-song melody of the playground to tell the story of a fickle "squaw" who resists nine eager suitors—and their offers of moccasins, feathers, and the like—before settling on "the tenth little Indian boy."

The nation's oldest college athletics conference is the Big Ten, but R&B sax master Bull Moose Jackson had a far different, far from officially sanctioned type of sport in mind on his signature "Big Ten Inch Record." The caesura that follows "ten-inch" is all that's needed to immortalize Fred Weismantel's kinda-dirty ditty as a classic of the double entendre. Aerosmith clearly knew what they were doing when they covered it in 1976. "Ten Seconds to Love," a premature ejaculation ode by Mötley Crüe, speaks to the same hormonally addled populace that Aerosmith tapped, only with a retooled message that it's OK to be bad in bed and brag about it afterward.

Presumably because of our 10 fingers, life in human society is set to base 10. We tend to lop things off in clusters of 10 and 100, and not just countable items, like apples and dollars, but even hypothetical constructs, like the 10-foot pole we wouldn't touch something with. We make Top 10 lists and rate people's looks on a scale of 1 to 10. When we feel fantastic, larger than life, we feel like XTC's "Ten Feet Tall," which features as concise a four-bar guitar solo as has ever been attempted. The acoustic, jazz chord-laden single from the band's watershed *Drums and Wires* was bassist Colin Moulding's attempt to subvert the band's quirky-jerky

---

1 According to Jimmy Page, it was while recording a version of Nilsson's "Ten Little Indians" with the Yardbirds that he invented the studio effect known as "reverse echo," which sounds pretty much like what its name implies: backwards-sounding and echo-laden. It appears on some heavy Zeppelin tracks, including "Whole Lotta Love" and "You Shook Me," but the Beatles song "I'm Only Sleeping," which came out a year earlier, employs a similar effect.

M.O. by writing something altogether smoother and sexier. "Ten Feet Tall" remains one of XTC's most delightful and understated creations. But don't take my word for it: Go ask Alice, when she's ten feet … tall.

The Stone Roses' "Ten Storey Love Song" amps up the love-as-height imagery to gargantuan levels. The sky-high *amor* of the title is in sync with the over-the-top ambitions of *Second Coming*, the Roses' swan song, which was all but universally reviled for its pompous title, bloated blooze riffs, and generally weak material. "Ten Storey Love Song" is one of the record's few standouts—tuneful and with a sense of proportion, even with its outsize emotions.[2]

"Box #10," Jim Croce's affecting tale of hard times in New York, refers to the address of the Sunday mission where the singer ends up after losing his earthly possessions to naiveté and a cold-hearted woman. That Sunday mission might plausibly be in the vicinity of Bruce Springsteen's famous "Tenth Avenue Freeze-Out," which traces the origins of the E. Street Band in colorful, if decidedly abstruse fashion. Clarence Clemons once admitted in an interview that he had no idea what it was about.

For a freeze-out with zero ambiguity, look no further than Gordon Lightfoot's "10 Degrees and Getting Colder," in which a down-on-his-luck musician attempts to hitchhike home to Milwaukee from the Boulder Dam. And on the subject of cold-weather pursuits, "Don't ski naked down Mount Everest/With lilies up your nose" advise the Monochrome Set, along with a litany of surrealistic matrimonial advice, in "Ten Don'ts for Honeymooners."

As a neat and handy measure, 10 crops up often in the context of the clock and the wallet. A $10 bill used to be called a sawbuck because of the Roman numeral X's resemblance to a certain wood-holding device. No one calls it that anymore, probably because it buys so little that it doesn't seem deserving of a jazzy nickname. Of course, M.I.A. wouldn't agree: In "$10," she declares, "What can I get for $10—anything you want," a sentiment that would go down well with ZZ Top's "Ten Dollar Man," from the transitional *Tejas*. Cheap Trick can lay claim to writing the definitive ode to 10:00 p.m., "Clock Strikes Ten," the final track from *At Budokan*. I like the Black Keys' "10 A.M. Automatic" for the morning slot, while the Verlaines' "All Joed Out" deserves a prize for containing the world's sole reference to "ten o'clock in the afternoon."

Ten has turned up frequently as an album title, on Pearl Jam's debut, for example, the record that brought grunge into America's living rooms, along with works by LL Cool J, the Smithereens, the Guess Who, Enuff Z'Nuff, Wet Wet Wet, and Asleep at the Wheel. The second discrete semiprime also fits into Sting's *Ten Summoner's Tales*, the Elvis Costello best-of *Ten Bloody Marys & Ten How's Your Fathers*, and has had featured roles in some killer tunes, like the 10 murdered oranges in John Cale's "Child's Christmas in Wales" and the 10-ton truck in that Smiths song.

But the very best of tens share a common denominator in the person of Jimmy Page. "Happenings Ten Years Time Ago" was the first Yardbirds single to feature their hot young guitarist, and it was a harbinger of (shapes of) things to come. Though it featured what would come to be seen as typical Page-ian/Zepploid bluster (Middle Eastern modalities, mystical lyrics, other psychedelic tropes), its nifty structure, shape-shifting production, and bevy of guitar sounds—stabbing, discordant, feedback-laden, explosive bursts, courtesy of Page and Jeff Beck—still amaze.

---

2 The Velvet Underground voiced a similar sentiment with the also-ran "Love Makes You Feel Ten Feet Tall," which ended up on *Loaded: The Fully Loaded Edition*.

If 6 Was 9

THE VERDICT

A wiser man would heed Dusty Springfield's "I Close My Eyes and Count to Ten," but my mind is made up. When I first began brainstorming ideas for this list at the bar that day, my #10 song came to me right away. And despite several excellent contenders (the Yardbirds song in particular is certainly epic enough to take the crown), I am still inclined to stick with Led Zeppelin's "Ten Years Gone." Here's a song that encapsulates all that is great about Zeppelin: the sense of space and majesty, indelible melodies, guitar lines that fly too close to the sun, drums that shake you to your foundation, and in this case, something like 14 separate guitar tracks during one especially rich sequence. From the converted there will be no argument. For those who never got into the band, or simply never got their appeal, for those who have come to hate "Stairway" or were born too late for Zeppelin to truly enter your soul, etc., I say unto you only this: This one might make a believer out of you, at least a believer in the song itself.

Like any great Led Zeppelin track, "Ten Years Gone" is an amazing feat, a miniature sound-only movie.[3] Every melodic excursion and turn within its six-minute confines sounds like it was written into the song, and yet there is a certain organic looseness that keeps it from coming off like the labored-over creation it clearly is. It starts hushed and builds elegantly upon an insistent, Möbius strip kind of a lick, one that keeps gaining strength as all the melodic permutations of it are writ large and strategically deployed. This paean to a lover who demanded he give up music for her is also one of Robert Plant's most modulated performances. When Percy finally delivers the payload and unleashes a couple of his signature *woo-woo, yea-y-yeah*s, like The Banshee incarnate, it's perfect, the only sound that will do.

A SONG TITLE	AN ALBUM	A BAND	A LYRIC
"Count to Ten" – The Clean (1982)	*X* – Kylie Minogue (2007)	10cc – Manchester, England, art rock (1970s–2010s)	"Ten silver saxes, a bass with a bow" – Neil Young, "Cinnamon Girl" (1970)

---

[3] "A movie theme without a movie" is how "Rotary 10" is described in the liner notes of R.E.M.'s odds-and-ends collection, *Dead Letter Office*.

11

Eleven is the redheaded stepchild of the first 20 natural numbers. If you need convincing, consider this: It took a fictional guitarist to put 11 on the rock 'n' roll map. Those specially made amps that go "one louder" represent rock's defining 11, and the battiness of Nigel Tufnel's contention underscores 11's problem. As *Spinal Tap*'s ostensible director Marty DiBergi sensibly suggests, why even *go* to 11? Why not just make 10 louder?

Eleven just doesn't get a lot of respect. It's forever in the shadow of 10. And fittingly, despite the not insubstantial number of #11 songs out there, precious few seem essential to a band's or an artist's canon. A significant number of 11s are unreleased songs, outtakes, live-only or some other aspect of less-than-top-tier status. U2's "Eleven O'Clock Tick Tock," for example, has remained in the band's set list since the '80s but never made it onto a studio album. A lilting hidden track tucked at the end of R.E.M's *Green* is known variously as "Untitled," "11," and "The Eleventh Untitled Song." Hüsker Dü's "Dozen Beats Eleven" is an outtake from *Zen Arcade*, while England's Doves inexplicably slapped the rousing "Eleven Miles Out" on a B-side.

The Grateful Dead's "The Eleven" fits the pattern. Recorded for the palindrome-titled *Aoxomoxoa*, the song's only official appearance was on the four-sided *Live/Dead*, a 1969 recording that caught the original lineup at the peak of its powers. Named for its circuitous 11/8 time signature (same as Primus's "Eleven"), "The Eleven" did not remain in the band's live repertory for very long; one suspects it was a bit too crazed and complex for the more laid-back, post-Pigpen Dead. When first developing this Phil Lesh-Robert Hunter creation, according to Jerry Garcia, the band would practice playing sequences of 11 relentlessly, a work ethic that might surprise those who associate the Dead with spontaneous improvisation and lugubrious, go-anywhere jams.

If the Grateful Dead were the Beatles of late-'60s San Francisco, then the West Coast Pop Art Experimental Band were the Freddie and the Dreamers of the scene. And their "Sweet Lady Eleven" would be just another forgettable track ("We'll travel to the moon/ We'll travel to my room! *Uuh!*") if not for the creepy notion that the 11 in the title seems to refer to the lady's *age*. It just might: *Where's My Daddy?*, the band's previous record, flaunted the Lolita lust of the group's troubled leader, Bob Markley. "Love on an 11-Year-Old Level," by Zappa protégés the GTOs, on the other hand, is merely banal. The GTO's ("Girls Together Outrageously") were the first, and hopefully the last "all-girl groupie group." While we're on the subject of youth, misspent or otherwise, the much-loved BBC comedy *The Young Ones* squeezed a live performance by a different band into every show. The first episode featured Nine Below Zero performing a rousing version of "Eleven Plus Eleven."

To be sure, some of the second unique prime's finest moments in the sun come as a lyric rather than as a title. Eleven, after all, is one of only two trisyllables under 20 (the other being 17). That may not sound like much, but several greats have made excellent use of its distinctive meter. Dylan's man in the coonskin cap in the big pen "wants eleven dollar bills, you only got ten," and David

Bowie's "Queen Bitch" is "up on the eleventh floor/watching the cruisers below." Michael Stipe liked the sound of 11 so much he mentioned it twice in the gorgeously morose "Perfect Circle," while Neil Young's "No Wonder" draws on the number's more recently acquired, darker associations ("That song from 9/11 keeps ringing through my head"). When Pete Townshend and Roger Daltrey played a memorial gig at Madison Square Garden a few months after 9/11, they were wise not to revisit the Who chestnut "It's Not True"—the lyric "I haven't got 11 kids/I weren't born in Baghdad" might have caused unintended controversy.

Something about 11 seems to stoke the fires of surrealism. The Bonzo Dog Band's "Eleven Mustachioed Daughters," from the surpassingly strange *The Doughnut in Granny's Greenhouse,* has the title characters "running in a field of fat" amid jungle percussion, snatches of party dialogue, and an extended coughing section. "Eleven Executioners," by the lascivious songwriter/provocateur Momus, is an intimate, waltz-time concoction that begins with "eleven gentlemen dancing" and proceeds to describe each of them committing murder of some sort, before a nonsense finale worthy of Brian Eno in pop mode. "VCL XI," a song by Orchestral Manoeuvres in the Dark, was also the name of the pre-OMD duo of Andy McCluskey and Paul Humphreys. Both song and band were named after a valve pictured on the back of Kraftwerk's *Radio-Activity*.

In the U.K., the late-morning repast known as elevenses usually means some tea and sweet cake, while in America, as Michael Pollan reminds us in *The Omnivore's Dilemma*, "the modern coffee break began as a late-morning whiskey break called the 'elevenses.'" For a three-song soundtrack to accompany your next elevenses, I recommend "Elevenses," a bebop instrumental by the renowned British jazz musician John Dankworth, followed by "Elevenses" by Neil Halstead of Slowdive, followed by the 50 seconds of space burps that make up "Elevenses" by FFWD. Then it's back to work.

Right now, before we take it all home, is the traditional moment for what's known in theater parlance as the eleven o'clock number. This is a showstopper that sets up the big windup. "Sit Down You're Rockin' the Boat" from *Guys and Dolls* is an archetypal eleven o'clock number, but for the life of me I can't find one with an 11 in the title. Anyway, enough 11th-hour dithering.

THE SCOTTISH CONNECTION:
"Just After 11 She Left" is a toy piano-driven number by the shockingly prolific Scottish singer-songwriter Kenny Anderson, aka King Creosote; "The Eleventh Earl of Mar," by Genesis, chronicles complex machinations in the life of 18th-century Scottish politician John Erskine. And Scotland's Highland and Islands Airports operates 11 regional airports. It's uncanny.

THE VERDICT

My date to the Eleven Ball is Blondie's "11:59," which kicks off side 2 of *Parallel Lines,* one of the great records of its era. Best known for the smooth disco of "Heart of Glass" and the hard crunch of "One Way or Another," *Parallel Lines* spans an array of musical flavors and nails them all. "11:59" is a Spectorish teenage symphony to God, a three-minute blast of new wave chug and girl-group melodrama that pulls out all the stops (and starts) along the way—the breakdown, the build-back-up, the strategic key change—for maximum impact. The band can barely contain itself, and Deborah Harry, finding the middle ground between the stylized silkiness of "Glass" and the tough-mama oomph of "One Way," lets loose alliterative lyrics with grit and plain-spoken urgency. Written by keyboard player Jimmy Destri, who didn't make it into the revived Blondie of the '90s and '00s, "11:59" begs to come blasting out of a jukebox in a steamy-windowed late-night bar, when time is running out—and you want to stay alive.

A SONG TITLE	AN ALBUM	A BAND	A LYRIC
"Home By Eleven" – Steve Alaimo (1959)	*11* – The Smithereens (1989)	Eleventh Dream Day – Chicago, alt-rock (1980s–2010s)	"When our pheromones are turned up to 11" – Magnetic Fields, "Time Enough For Rocking When We're Old" (1999)

POSTSCRIPT: "11:59 (It's January)" is a dark New Year's Eve tale of counting down to midnight and projecting beyond into post-holiday depression. Scrawl, a trio from Columbus, Ohio, were active in the mid-'80s male-centric indie rock scene and have made some powerful, eloquent records that stand their ground remarkably well.

12

Twelve marks a distinct departure from each of the positive integers that precedes it, for one simple reason: It doesn't rhyme with much of anything. Yes, there's "delve" and "shelve," but neither has been of much use to lyricists.[1] On "Three and Nine," an especially mathematically minded concoction from Roxy Music's *Country Life*, Bryan Ferry croons, "Whether making out or played out/Three and nine make twelve." But wisely he doesn't attempt a rhyme. Deeply un-euphonious—that's 12, which is why people prefer to nickname it. A set of twelve things is a dozen or a gross (or a duodecad, if you want to get technical); 12:00 is noon or midnight. Although in "Atlantis," Donovan found something eloquent in the number ("On board were the Twelve: the poet, the physician, the farmer, the scientist, the magician and other so-called gods of our legends"), you get the feeling that most people just find the word ungainly.

Yet 12 is simply too essential to our existence to not find its way into song titles. Certainly the elephant in the room is "The Twelve Days of Christmas," which dates to the late 1700s and has been recorded hundreds of times with innumerable variations. It takes a rare type of universality to appeal to Pat Boone, Insane Clown Posse, and Belle & Sebastian.

Twelve lends itself to clock-quoting lyrics like no other number, and almost exclusively in a nighttime sense. When Lucille Bogan sings, "And if you want my meat/You can come to my house at twelve" in her bawdy 1935 blues song "Barbecue Bess," she definitely doesn't mean noon (or meat, for that matter). Bogan's "Black Angel Blues" begins "I've got a sweet black angel," a line that opens a similarly titled song from the Stones' *Exile on Main Street* that's credited to Jagger/Richard. Mick Jagger, by the way, surely means 12 a.m. in "Miss You" when he says he'll come 'round at twelve with some Puerto Rican girls that's just *dyin' to mee-chew*. Late-night plans are afoot in the Strokes' "12:51," only where Jagger says he'll bring a case of wine, Julian Casablancas is cool with just picking up a couple 40s and skipping the party. In "Dirty Water," which enjoys eternal life on oldies radio, the L.A.-based Standells mocked the curfew policy of 1966-era Boston University ("Frustrated women have to be in by 12 o'clock"). "Twelve O'Clock Tonight" finds Doris Day plotting a midnight tryst, albeit a quiet one, so as not to wake her folks, while Bo Diddley, who declared in "The Story of Bo Diddley" that he was "born one night about twelve o'clock," sounds like he's on a late-night prowl on the fiddle-dominated instrumental "Clock Strikes Twelve." Little Walter, master of the blues harmonica, finds a similar panther-walking pace in "Quarter to Twelve."

The clock is just one of 12's major claims to fame. The 12-inch maxi-single, first made available to the record-buying public in 1976 with offerings like "Love to Love You Baby" and "Theme From 'Shaft,'" was an industry standard until CD singles and promo videos supplanted it. Nevertheless, the format remains a favorite of DJs, crate diggers, and New Order fans. The 12-bar blues and its three chords provide the basis for every blues song ever written, and 12-bar-based songs like "Johnny B. Goode" and "Rocket 88" pretty much defined rock in its infancy. Tunes with titles like "12 Bar Blues" (NRBQ) and "12 Bar Original" (initially unreleased by

---

1 In the Drive-By Truckers' "Go-Go Boots," Patterson Hood gets around the issue with a close-enough rhyme: "Daddy's been preachin' the word since he was 12/all about the merciful savior and the fires of hell."

the Beatles) do come to mind, but with rare exceptions, like Richard X. Heyman's cautionary "Twelve Bars and I Still Have the Blues," the 12-bar structure is mostly an implicit construct.

Many a rocker has enlisted the help of a 12-step program. Jason Pierce, the irascible sonic explorer behind Spiritualized who also goes by the name J. Spaceman, plunders the program for imagery in "The Twelve Steps," a hard-charging riff rocker that owes debts to such '70s-grade bombast as Deep Purple's "Highway Star" and the Moody Blues' "The Story in Your Eyes." Slathered in layers of guitars and augmented with police sirens, wailing harmonica, and orchestral strings, the song charges forward like a junkie to his stash. And while the lyrics are often submerged beneath the maelstrom, Pierce makes sure to enunciate step 8 ("Fuck being straight") in his decidedly non-Bill W-approved sequence. "Twelve Steps to Love," a 1964 single by the Tremeloes, offers a more traditional set of directives ("11th step/meet her friends and relations/12th step/sendin' out invitations").

AWARD FOR INNOVATION IN #12: In the bedroom procedural "12 Play," R. Kelly trebles the concept of foreplay for a generation that just seems to need more.

The clarion harmonies of the Mamas and the Papas once counted among life's simple joys. But, for me at any rate, the Papa John Phillips incest allegations of 2009 have rendered the enjoyment of his songs a far more complex proposition. No doubt, "12:30 (Young Girls Are Coming to the Canyon)" is easy to swoon over. Like the springtime version of "California Dreamin'," "12:30" posits the Golden State as an antidote to the Big Apple, which it depicts as a place so unrighteous that no one can even be bothered to fix the busted clock atop the church steeple. While I admire that detail, there's one line that still makes me recoil a bit when I hear it, and not because of Phillips's personal history, but because it is simply heinous: "At first so strange to feel so friendly/to say good morning and really mean it." Think about every bright-eyed hippie in Laurel Canyon saying good morning to you—and *really* meaning it. And that subtitle (Young Girls Are Coming to the Canyon) raises questions too: How young? Is "canyon" a metaphor? Et cetera.

It seems all but a mathematical certainty that any given number will appear *somewhere* in Bob Dylan's sprawling songbook, and 12 is no exception.[2] In "Rainy Day Women 12 & 35," 12's meaning is oblique, a mere ornament in the title, but in "A Hard Rain's A-Gonna Fall" Dylan gives 12 its due with the toothsome line, "I stumbled on the side of twelve misty mountains." "Twelve Gates to the City" is a traditional gospel song whose chorus, "Oh, what a beautiful city," was trilled by Joan Baez and many others during the early-'60s folk boom, but give me Blind Boy Fuller and Sonny Terry's rustic take or the version by the Rev. Gary Davis, which is arguably the definitive one. On that religious note, it would be sacrilege not to mention the twelve disciples, Jesus' first followers (pre-Twitter), who inadvertently spawned musical acts ranging from the Young Disciples and the Lost Disciples to the Disciples of, oh, just about anything—death, power, smoke, sound, and my favorite: the Disciples of Torak.

There's no 12 in Dylan's version of "Girl From the North Country," but British folk troubadour Roy Harper has long performed the song, often with an outrageous faux-Dylan delivery. I bring up Roy Harper because he has two gorgeous works of duodenary folk-pop to his credit. Perhaps best known to American audiences for his lead vocal on Pink Floyd's "Have a Cigar," Harper is an influential behind-the-scenes singer-songwriter who coaxed glorious acoustic solos from Jimmy Page[3] on the 1971 LP *Stormcock*

---

2 Dylan's canon includes mentions of zero through 17—surely a record.
3 Mr. Page was always a huge Harper fan; see "Hats Off to (Roy) Harper" from *Led Zeppelin III*.

and has amassed a fascinating if erratic body of work in ensuing decades. "Twelve Hours of Sunset," as shimmery as its title would suggest, is an almost meterless meditation gilded with spare acoustic strums evoking the title's "twilight sublime." Roy's other #12, "October 12," from his 1966 debut, *Sophisticated Beggar*, is a despairing tale told by a man who has left the earth. Over restless fingerpicking, in imagery that's both familiar and unfathomable, Harper ponders religion, self-delusion, the arbitrariness of life, and other dark topics. He delivers the refrain, "And I hated myself when I was living," as if there's no inherent contradiction.

While Harper was keen to consider time and timelessness, he had the good sense not to cover "The Twelfth of Never," a cauldron of frothy cheese that borrows the tune from "I Gave My Lover a Cherry" (the song that got the earnest strummer's axe smashed in *Animal House*) and turns it into world-class schmaltz. Johnny Mathis popularized it in 1957; Cliff Richard redid it in 1964; Donny Osmond scored with it in 1976; hell, Jeff Buckley was known to cover it. But the song's inherent sappiness simply cannot be outrun. Too bad Earth Wind & Fire's "Fantasy" was not named for the phrase on which it leans so heavily; then you could just refer to the good "Twelfth of Never" and the bad one.

THE VERDICT

"Rate my love on a scale of 1 to 10, and I'm sure you'd give it 12," sang the SOS Band on the irresistibly slinky "No One's Gonna Love You." Echoing that sentiment, on a scale of 1 to 10, I would give a 12 to "12 Red Roses" by Betty Harris, a world-class soul singer whose reputation among true aficionados far exceeds her record sales or name recognition.

Kicking off with a snare drum tattoo that sounds uncannily like the one Charlie Watts used to open "Get Off of My Cloud" (both songs were recorded in 1965), the song recounts the ups and downs of a love affair. The second rose "reminds me of the good times," the ninth recalls "the 9 out of 10 times you didn't come through." And the 12th is all about the future.

Like the majority of her songs, "12 Red Roses" was written and produced by Allen Toussaint, the living embodiment of New Orleans musical tradition, and with its laid-back but supremely funky rhythm section, Toussaint's piano, and a masterful vocal by Ms. Harris, the song has all the makings of a classic soul nugget. But somehow it never even made the U.S. charts.

Best known for her steamy cover of Solomon Burke's "Cry to Me," Betty Harris recorded 28 songs between '63 and '70 before leaving the music business behind. But she made an unlikely return to recording in 2007, partly inspired by website tributes from thousands of fans touting her as one of the greats. Though never a diva, she has not been averse to a few accolades. In a 2004 interview, Ms. Harris put it this way: "[T]hose who label me the true Soul Queen of New Orleans, you are well within your rights."

A SONG TITLE	AN ALBUM	A BAND	A LYRIC
"Twelfth House" – New Musik (1982)	*Twelve Dreams of Dr. Sardonicus* – Spirit (1970)	Twelve Clouds of Joy – Kansas City, big band/swing (1930s)	"I was dancing when I was 12" – T Rex, "Cosmic Dancer" (1971)

13

Fear of the number 13 is an ancient phobia. Whether it dates to the time of Hammurabi or Jesus, the Vikings or the Knights Templar, *triskaidekaphobia* is a symptom of the human condition. We can't bring ourselves to call the 13th floor of a building by its proper name, and we still get a funny feeling on Friday the 13th. Fear of that particular day, or *friggatriskaidekaphobia*, may not be quite as ancient, but it too beclouds the minds of many, even in this putatively modern age. Songwriters haven't exactly avoided 13, but plenty have steered clear. The closest Dylan comes to it is the mention of an Italian poet from the 13th century, whose book of poems makes a cameo appearance in verse 5 of "Tangled Up in Blue." Tom Petty is at his most Dylanesque taking on J. J. Cale's road song "Thirteen Days," which contains the built-in concert applause line, "We're smoking cigarettes and reefer/drinking coffee and booze."

Glenn Danzig is not usually mentioned in the same breath as folks like Dylan and Petty. He's considered a metal behemoth first and a songwriter second, yet with "Thirteen," Danzig joined the rarified company of Bob and Tom by having his song covered by Johnny Cash. Johnny's version of Danzig's "Thirteen," just him and a guitar, is arguably more truly badass than Glenn's full-on treatment, but give credit where it's due: "Thirteen" is a starkly powerful song, and Cash interprets it with biblical certainty.

One has to wonder, though, how the agelessly acerbic Elvis Costello felt about all this. A few years earlier, Elvis had written "Complicated Shadows" for Mr. Cash, but the Man in Black had passed on it. Now, you'd have to figure the smart money was on Costello, not Danzig, to write a song that Johnny Cash would cover. Presumably, though, even in writing a "Johnny Cash song," Elvis could not help but leave his complicated fingerprints all over it. In a tune he wrote for a different singer, Wendy James, Costello couldn't resist the line "I want to stand forever/From now 'till the thirteenth of never." Happily, Elvis nailed 13 once and for all in "13 Steps Lead Down" from *Brutal Youth*, his reunion with the Attractions after an almost decade-long immersion in a stew of non-rock styles.

The early rock 'n' roll era serves up a few choice 13s. Chuck Berry's Latin-tinged "Thirteen Question Method" is a vintage numerical procedural. But tragically (for me, at any rate), Chuck's method is on the same shaky theoretical ground as Paul Simon's "50 Ways to Leave Your Lover." Just as Simon delivers a mere four ways to leave your lover, Berry's 13-question method includes roughly eight actual questions. Question No. 1 ("let's have some fun"), for example, is not a question. Neither is question No. 5 ("I won't give you no jive"). It's surprising that as sharp a lyricist as Berry could abandon the song's premise so completely. Still, aside from this egregious lapse in logic, it's a delightful bonbon. Ry Cooder does a sweet cover of it.

In "Thirteen Women," jazz guitarist Dickie Thompson dared to posit the sunny side of being nuked: nookie. It might seem crass today, but if you can just accept the idea that a man could imagine the unthinkable and still like his odds, this song is for you. The version made famous by Bill Haley & His Comets was supposed to be a single, but after "Rock Around the Clock" played over the opening credits of *Blackboard Jungle* in 1955, the powers that be opted for the rocker as the A-side. Half a century later, it still seems like a wise decision, even though "Thirteen Women" is catchy as all get-out. Ann-Margret's version of the song, "Thirteen Men," steals a page from Peggy Lee's finger-popping "Fever," while the Renegades—British garage-punkers who favored U.S. Civil War garb—played "Thirteen Women" as a cave stomp with tacky organ and fuzzbox turned up to 11.[1]

Not many people ever got to see the Velvet Underground perform live, but, as Brian Eno memorably put it, everyone who did went on to start a band. The same could be said about the Monochrome Set, except the part about starting a band. The Velvets, who added seedy to rock's vocabulary at the height of the patchouli era, still inspire people today, but conjuring anything approximating the sound of the Monochrome Set—like Noel Coward singing surf music—would be nigh impossible. Yet the Set's literate, strummy pop and cheeky sensibility can be felt in the work of early acolytes, like Morrissey, Marr, and Edwyn Collins, and more recently Franz Ferdinand.

Wes Anderson would be wise to use a Monochrome Set song in one of his movies. "On the 13th Day," for example, from 1982's *Eligible Bachelors*, might work, only he'd do well to skip the angular refrain and instead thread the song's entrancing and unexpected Byrds-like break into a scene where, say, Owen Wilson tinkers with his childhood electric train set, or Jason Schwartzman irons a cricket outfit.

While Wes Anderson is roundly worshiped for the ingenuity of his soundtrack choices, John Carpenter stands at the apex of musical achievement by a film director. His piano theme from *Halloween* remains indelible shorthand for cinematic dread, and his score for *Assault on Precinct 13* has earned more praise than the film itself, and the film isn't bad at all. The main title—equal parts Zeppelin ("The Immigrant Song") and Schifrin (film composer Lalo, an immigrant from Buenos Aires)—is especially tasty.

XIII

THIS IS 13

The comeback album by Anvil contains 13 songs (and bonus track "Thumb Hang"). *13*, Black Sabbath's return to its almost-original lineup in 2013, had eight tracks, but to the faithful, eight was enough.

"It was a bright cold day in April, and the clocks were striking thirteen."
—*1984*, George Orwell

Nothing says off-kilter like 13. Witness the Munsters' 1313 Mockingbird Lane address, or Francis Ford Coppola's film debut, *Dementia 13*. Garage rock compilations are full of songs like "13 O'Clock (Theme For Psychotics)" by Mouse and the Traps, "13 Stories High" by Tonto & the Renegades, and the queasily trippy "Thirteen O'Clock Flight to Psychedelphia" by Plato & the Philosophers—all of which connect 13 with losing your mind.

I will confess to experiencing a bit of psychosis of my own in choosing a winner for this slot. How to choose from the musical apples and oranges vying for #13 gold? Certainly Johnny Cash's "Thirteen" is worthy, embodying, as it does, 13's essential bad mojo so completely. It's like the evil twin to another major contender, Big Star's "Thirteen"—the tenderest song imaginable and one of the best loved by a much-loved band. Both last about two and a half minutes; both are just a singer accompanied by a guitar (although Big Star's mix is embellished with phased harmony vocals). Cash's subject is a man damned to a living hell, while Big

---

1   *Thirteen Women* is also the name of a Depression-era film that contains the only movie appearance of Peg Entwistle (no relation to bassist John of the Who), who became the stuff of sad legend when she leapt to her death from the HOLLYWOOD sign in 1932.

Star's song is about getting to first base. Then there's the Pixies' scabrous "No. 13 Baby." It has the quintessential Pixies ebb-flow between crunch and swoon, and the number is front and center, the star of the show. Even the cover art for *Doolittle*, whence it came, is strewn with low numbers. Surely, any one of these three is worthy of claiming the haunted numeral.

THE VERDICT

Faced with the choice of Big Star's sweet apple, the Pixies' spicy *naranja*, and Cash/Danzig's devil's fruit, I'm going with the apple. Chilton's "Thirteen" is not dangerous or sexy; it wears its heart on its sleeve, like the smitten kids in the song. But the song is not devoid of darkness. "Won't you tell your dad get off my back," one of them sings in verse 2, followed by the wonderfully cryptic, "Tell him what we said 'bout 'Paint It Black.'"[2]

We'll never know exactly what they said about "Paint It Black." But that little secret adds to the magic of "Thirteen." And the numerous cover versions it's inspired—whether lovingly faithful (Elliott Smith, Wilco, the Lemonheads) or wholly misguided (Garbage, Counting Crows)—have always struck me as making a strong collective argument for leaving "Thirteen" undisturbed.[3] That is, until I heard a version by Skylar Gudasz, a North Carolina-based singer-songwriter who's been handling "Thirteen" duties, and earning raves, for a few years now as part of Big Star Third Live. This unlikely orchestral reenactment of the band's dark swan song has traveled the world since 2011, involving any number of Big Star acolytes along the way and featuring Jody Stephens, the lone surviving band member, on drums.

When I spoke with Skylar Gudasz, I had to ask what she thought they said about "Paint It Black." Here's what she said: "The way I think about it is these two punk kids. The dad is like, 'Ah yeah, tell that dude don't come around here with his rock 'n' roll! And the kid's like, 'No, no, no – 'Paint It Black'! That's the one. It's a great song!'"

A SONG TITLE	AN ALBUM	A BAND	A LYRIC
"13 (Under a Bad Sign)" – Sloan (2014)	*13* – Lee Hazlewood (1972)	The 13th Floor Elevators – Austin, Texas, psychedelic garage rock (1960s)	"I lived with thirteen dead cats/ a purple dog who wears spats" – The Velvet Underground, "I Can't Stand It" (1969)

POSTSCRIPT: The Rolling Stones released "Paint It Black" as a single in May 1966—*on Friday the 13th.*

---

2 Or is it "Paint It, Black"? That comma, apparently added by Decca Records after initial pressings of the single, ranks among the more puzzling bits of editorial imprecision in rock (right up there with "the Mama's and the Papa's" and the Zombies' *Odessey and Oracle*). Most digital versions have reverted to the simpler, comma-less title, but iTunes does offer Original Spelling. The one with the comma also gives you a slightly longer fadeout. Your call.
3 No has clamored to cover "No. 13 Baby," likely because, as with most Pixies songs, it amounts to a vocal decathlon. As for Danzig's "Thirteen," who would want to cover a song that Johnny Cash owns?

> "Are you about a size 14?"
> —Jame Gumb, aka "Buffalo Bill," *The Silence of the Lambs*

Singing the praises of 14-year-old girls is a proud rock tradition that unites such disparate personages as Bob Weir of the Grateful Dead ("So instead I've got a bottle/and a girl who's just 14")[1] and Iggy Pop ("My girlfriend Betsy she's just 14").[2] Not just any 14-year-old girl, mind you—Bob and Iggy were singing about the emancipated type. For 14 is the youngest acceptable age in the U.S. at which a youth can be deemed an "emancipated minor," entitled by law to operate under no one's authority but his or her own. In "Superlungs My Supergirl," Donovan extolled the virtues of one such emancipated minor—not her romantic allure but her power to inhale ("She's only 14 but she knows how to draw").[3] Terry Reid, in his version, howled Donovan's original lyrics ("She's too busy getting high with her classmates in school") with gusto. But when Mr. Leitch released the song in a cleaned-up version, Superlungs had implausibly ditched grass for watercolors—now she's too busy "painting *sky* with her classmates in school." Please. Emancipated minors have no interest in painting sky.

But seriously, is 14 the first semi-pointless number? At first glance it might seem so. The first 10 are like supernumerary gods. From there, yes, 11 lacks a clear personality, but it rhymes with "seven" and thus is pretty essential. Twelve has the clock and the months. Thirteen has the spooky bad-luck thing. But 14? Not many inherent associations, unless you count the "14 joys and a will to be merry" in Fleetwood Mac's "Sentimental Lady" (truly one of the more nonsensical lyrics in the Mac songbook). Gene Watson worked 14's gold connection with "Fourteen Carat Mind," a No. 1 country chart-topper in 1982. Valentine's Day falls on the 14th of February, inspiring Drive-By Truckers' cry-in-your-beer-worthy "February 14." Similarly minded is "February Fourteenth" by the Lilys, extremely competent My Bloody Valentine acolytes. And lest we forget: the Who's immortal *Live at Leeds* concert took place on Feb. 14, 1970. But the number might just as well mean anything. In Chris Stamey's "14 Shades of Green," for example, the singer yearns to see his lover's face turn the subtle gradations of color alluded to in the song title.

Despite 14's lack of zing in the song-naming department, it has a surprisingly rich pedigree. Several musical giants have employed 14 in their work. Dylan again: "In fourteen months I've only smiled once, and I didn't do it consciously" is a cracking

---

1 "Mexicali Blues," originally released on Weir's solo, Dead-abetted *Ace* (1972).
2 "Dog Food," from *Soldier* (1980).
3 And it was not Donovan's only extolment of feminine 14-hood. On one of his biggest hits, "Mellow Yellow," he sang, "I'm just mad about 14/she's just mad about me." We were living in a different age; it was deemed sorta-kinda acceptable for a man in his 20s to brag about his lust for underage girls.

line from "Up to Me," a *Blood On the Tracks* outtake that conflates "Simple Twist of Fate" and "Shelter From the Storm." One of Rod Stewart's finest creations, "Mandolin Wind," poignantly evokes "the coldest winter in almost fourteen years," while Neil Young fairly leans into 14 in the opening lines of "Time Fades Away" ("Fourteen junkies, too weak to work"), the lead and title track of a loose, wired-sounding set of live performances from 1973. Over the years, Young has disparaged *Time Fades Away*, and it took until 2014 for it to be re-released. North of Neil's Pain Street, way up on the avenue of trees, "Cypress Avenue" is where Van Morrison sings, "So young and bold, fourteen years old …" Back in the vicinity of Pain Street, the Clash's non-basketball-related "Washington Bullets" goes, "Oh, Mama, Mama, look there/a youth of 14 got shot down there."

Given 14's workaday vibe, it feels apt that purveyors of a certain lo-fi aesthetic seem to have embraced the numeral. "14 Cheerleader Coldfront" has the rough beauty of archetypal Guided By Voices, even if it's sung by Tobin Sprout, the band's sometime second vocalist/songwriter, whose name could almost pass for a GBV song title. (Similarly, the lead vocalist on the Guns N' Roses song "14 Years" is not Axl Rose but second-banana rhythm guitarist, Izzy Stradlin.) In a blistering, high-kicking performance during the 2011 Hopscotch Music Festival in Raleigh, N.C., Robert Pollard recounted the song's origins: "We were at a football game, and there was this line of 14 cheerleaders—14 cheerleaders all cheering for no one but themselves."

As with the Woodstock Festival, the 14-Hour Technicolour Dream—a 1967 happening at London's Alexandra Palace—was commemorated in song. Perhaps unsurprisingly, the less remembered festival inspired the less memorable song, in this case by The Syn, whose original bassist was none other than future caped Yes man Chris Squire. While Joni Mitchell's "Woodstock" achieves poetry with its "bomber death planes … turning into butterflies above our nation," "14 Hour Technicolour Dream" is far more prosaic. Nevertheless, the exhortations of Syn vocalist Steve Nardelli during the wind-out are worth quoting: "Suzi Creamcheese gonna be there, yeah!/Have a cabana, and smoke a banana if you want to." (Here's another banana reference: Brave Combo's goofball "Fourteen," which mentions "14 tons of golden ripe bananas" and will sound familiar to anyone versed in the XTC song "Scissor Man.")

New York's 14th Street[4] is celebrated for its seediness, kind of like Naples. Blind Joe Taggart's "14th Street Blues" laments the avenue's women, concluding, "I feel like snapping my big gun in your face/had the nerve to tell me 'nother man's got my place." For Rufus Wainwright, the road is the setting for heartbreak in his mournful "14th Street," and it's the site of the epic dance-off depicted in Gamble and Huff's "There's Gonna Be a Showdown," a barnburner first done by Archie Bell & the Drells and later covered by the New York Dolls. When Sylvain Sylvain launched his brief post-Dolls solo career, his first single was "14th Street Beat," a bouncy rave-up replete with the door-closing "bing-bong."

THE VERDICT

Television Personalities were shambolic before your favorite band was shambolic. The band's first single, "14th Floor," is what Chuck Berry would have written if he had been a young Brit living in council housing in 1977. In a weary voice, Dan Treacy echoes Johnny Rotten's "no future," not as a shouted screed accompanied by an upraised middle finger but as a half-sung shrug and a sigh.

---

4   It would not be surprising to encounter a wood louse on 14th Street—a wood louse, by the way, has 14 legs.

"14th Floor" certainly had the DIY punk ethos: Treacy and his compatriots performed the song live in the studio for a total cost of £400. They were not even aware that multi-track technology gave them the option of playing the music first and singing over it afterward. Treacy has had several emotional breakdowns in his rough-and-tumble career, the first one at age 14. He attributed it to being bullied. But, Treacy extrapolated, "At age 14, every boy is schizophrenic. It's hormones."

A SONG TITLE	AN ALBUM	A BAND	A LYRIC
"West 14" – Gol Gappas (1986)	*14 Songs* – Paul Westerberg (1993)	14 Iced Bears – England post-punk (1980s–1990s)	"Although she's only fourteen/ she really knows her courting" – Squeeze, "Vicky Verky" (1980)

15

Andy Warhol, who famously proclaimed that in the future everyone would be famous for 15 minutes, probably kicked himself for not taking ownership of that phrase. But apparently, once uttered, it was uncopyrightable. "Uncopyrightable," by the way, contains 15 letters and doesn't repeat one, making it (and a few others, e.g., "hydropneumatics") the longest such word in the English language.

David Bowie (who wrote a song called "Andy Warhol") could have copyrighted "TVC 15" for a line of holographic televisions like the one in the song of the same name. Surely, if the technology's not available now, it will be soon. But as for "TVC 15," it can't really be deemed a proper #15 song, since the digits are pronounced "one five." Copyright issues probably never entered the collective mind of Eater, who compressed Alice Cooper's "I'm Eighteen" into a 67-second eruption and called it "Fifteen." But given that the song's title was also the average age of these first-wave punkers, the breach seems forgivable. Warhol might have even been impressed by the young Londoners' act of aggressive appropriation.

Songwriters have responded to Warhol's 15 minutes of fame concept, early and often. "Warhol's 15," by the Blue Aeroplanes of Bristol, England—that rare band to actually get somewhere with a vocalist who simply spoke his lines as a poet would—has a lot in common with a Neil Young song, except for lyrics like "Your face is the future/ a smoother transaction is hoped for/ than the hollows beneath their skin." Warhol's notion crosses all borders. Songs called "15 Minutes" or "Fifteen Minutes of Fame" unite early glam queen Suzi Quatro, '90s house proponents Sheep on Drugs, Liverpool indie rockers Johnny Boy, industrial rockers Gravity Kills, and the Nugent vehicle Damn Yankees under a common Pop Art umbrella. In Kirsty MacColl's "Fifteen Minutes," the sorely missed singer dashed off a charming, Kinks-ish hate note to sellouts, bozos, and others "whose mediocrity excels."[1]

Fifteen years doesn't share the same cachet as 15 minutes, obviously, nor does 15 months or 15 seconds. And many of these songs are, quite frankly, bummers, like the mournful "Fifteen Years Ago," a country standard sung by Conway Twitty, Charlie Pride, and the Statler Brothers, about the gal who got away. "Fifteen Years" by the Levellers is about her too. "Fifteen Months" by Joan Baez chronicles a woman's solitude as she awaits her man's release from prison, while bucking the trend is "Fifteen Seconds," a swoon-worthy slice of strummy lounge-pop by the short-lived Ivy, featuring the airy, French-inflected vocals of Dominique Durand.

The title of Nick Cave's "Fifteen Feet of Pure White Snow" refers to the variety of white powder that sometimes gets cut with baby laxative. Do not confuse it with "15 Feet of Snow" by another Australian, Johnny Diesel—here, the snow is simply where Diesel will "go-o-ho/just to feel your glow-aho." The Mountain Goats memorialize reggae's Prince Far I in "September 15, 1983,"[2] while another date song, "February 15th, Happy Birthday to Me" by Bright Eyes, is a boozy lament for wasted years.

---

[1] Her chum Billy Bragg refers alliteratively to "the fifteen fame-filled minutes of the fanzine writer" in "The Great Leap Forward."
[2] The eminent DJ, producer, and musician died in a robbery at his home on the day referenced in the song title. Peter Tosh died in a home invasion four years later.

Long before the Parents Music Resource Center named its "Filthy Fifteen" most objectionable pop songs circa 1985 (No. 1 was Prince's "Darling Nikki" followed by Sheena Easton's "Sugar Walls"), carnal musical tributes to 15-year-old girls were fairly commonplace in the raunchier precincts of rock and blues. "I can see that you're 15 years old," goes the Stones' "Stray Cat Blues," "no I don't want your ID." Harry Belafonte, on the other hand, was extremely chaste. The love object in his song "Fifteen" is "the sweetest wine in the world." In modern times, the essence of 15-year-old girlhood was well summed up by an actual girl of 15 or thereabouts, Taylor Swift, whose G-rated "Fifteen" efficiently details the classroom crush, the first kiss, and the inevitable tragic breakup. For a more nuanced view of tragedy, try Rilo Kiley's "15," a tale of youth corrupted ("How could he have known she was only 15?") and Depeche Mode's mournful "Little 15," about a woman's doomed affair with a teenager who helps her "forget the world outside."

When Radiohead made *In Rainbows* available online on a pay-what-you-want basis, 62% of downloaders opted to pay nothing. Regardless, the record was an unqualified success, financially and artistically. The complex opener, "15 Step," which Thom Yorke said drew inspiration both from Brubeck's "Take Five" and Peaches' "Fuck the Pain Away," is a stand-out, with Yorke spinning out a vocal as only he can, threading a line over the song's sputtering 5/4 meter, brocaded by Jonny Greenwood's spare, elegant guitar commentary.[3] Elvis Costello's "15 Petals" is similarly complex, although its amalgamation of Latin rhythms, *West Side Story*-like brass discordance, and Klezmer touches, combined with Mr. C's typically byzantine verbiage, feels a bit overstuffed.

In Wire's "The 15th," the music trumps literal meaning at no expense to the total package. Why call it the 15th? The numerical name of the album it came from, *154*, makes sense (154 was the number of gigs the band had played up to that point), but the number 15 does not appear in the song. What the lyrics do suggest is some unfathomable inevitability, some unnamed thing forced into being by a different unnamed thing, and then forced out of existence: "Providing, deciding, it was soon there/Squared to it, faced to it, it was not there."

But it really doesn't matter. The oblique words just fit. The song is right. A steady pulse of distorted guitar, Colin Newman's tenor vocals—English choirboy gene-spliced with cyborg—paint a rich sound picture, with Wire's singular icy majesty in bold relief. FischerSpooner's cover version is damned good, while the Southern-accented vocals on Mike Watt's take, from 1996's *Whore: Tribute to Wire*, take a bit of getting used to. Beck has played it live, as did the Jay Reatard side project Angry Angles. But like many things, including icy majesty, the original is best.

THE VERDICT

In this book's initial printing, my choice for #15 honors was the aforementioned Wire song. From the start, it seemed self-evident that songs that use a number in the title with precision hold more currency than those that don't. "The 15th" never mentions the number itself, but somehow the way the title proclaims its own definitiveness was something I found deeply compelling. Next to it, the Who's "5:15," with the 15 coming after the 5, looked rather anemic in the #15 department. Eventually, though, that criterion proved utterly unworkable. Moreover, my initial exclusion of "5:15," one of my favorite songs by one of the great bands, was baseless

---

[3] Not to mention the finest string of etceteras this side of the Smiths' "Sweet and Tender Hooligan."

anyway. For while 15[4] is admittedly rather tangential in the title, the numeral is used in a highly specific way as a lyric—and on the first line of the song:

"Girls of fifteen," Roger Daltrey sings out with gusto.

"*Sexually knowing*" comes the joyous, harmonized response.

It's a deliberately disjointed image in a song stuffed with them ("He-man drag," "grayly outrageous"), and it's the kind of line only Pete Townshend would enunciate. The Beatles may have written about *women* who were sexually knowing—think of "Drive My Car"—but imagine how "I Saw Her Standing There" would have come across with an opening line of "She was just *fifteen/* and you know what I mean." The Stones, of course, had no such compunction in "Stray Cat Blues," where no bones are made about a 15-year-old as sexual pray for the salacious singer. But the teen girls in "5:15" are a breed apart. They know something crucial that Jimmy, the record's pill-addled protagonist, has yet to learn.

All theorizing aside, "5:15" is just a monster Who song, with a brawny, call-and-response-structured brass riff that swaggers like *Sticky Fingers*-era Stones. Released as a single in September 1973, it still feels like an album track, one that serves as a key moment on *Quadrophenia*. Yet, as Daltrey later said, no other song on the record could have passed muster as a single. The song consists of nothing but beautiful parts, and at any point one can simply marvel at the unpredictable thunder of Keith Moon's drums and John Entwistle's miraculous bass playing. It's bookended by the succinct and perfect summation of any teen's turmoil: "Why should I care? Why should I care?" And consider this: "5:15" contains what has to be the third best scream in Who history. Obviously the acrobatic wail that climaxes "Won't Get Fooled Again" is No. 1. Daltrey's work in "Love Reign O'er Me" is the silver medalist in the great Who Scream Olympics; but surely the "*wow!*" that Roger emits in his final vamp on "5:15," right after *"Out of my brain/on the train on the train,"* takes the bronze. It's bantam weight compared with the other two, yet so well timed and pitch-perfect as to be worthy of scrutiny, wonder, and imitation for as long as there is a rock to roll.

A SONG TITLE	AN ALBUM	A BAND	A LYRIC
"Fifteen Flies in the Marmalade" – Legendary Pink Dots (1987)	*15 Big Ones* – The Beach Boys (1976)	Fifteen – Bay Area, punk revival (1990s-2000s)	"Fifteen was chosen because he was dumb/ seven because he was blind" – Brian Eno, "Back in Judy's Jungle" (1974)

---

[4] It's not the only instance of 15 on *Quadrophenia*. On "Sea and Sand" Jimmy sings, "Should have split home at fifteen."

16

The birthdays of girls turning 16 are celebrated across cultures, but it's the American tradition of the Sweet 16 party that's engendered its own commercially successful song genre. Notwithstanding the 16 candles burning on Al Green's wall in "Take Me to the River" (turning him into the biggest fool of them all), most enduring songs of this ilk tend to be pretty sweet and innocent.[1] The Crests' "16 Candles," voiced by the operatic Johnny Maestro (formerly Mastrangelo), is a quintessential high school slow dance with an inspired doo-wop vocal arrangement. "You're the prettiest, loveliest girl I've ever seen," croons Johnny, his voice nearly breaking at the upper reaches, but whether the object of his affections ever returns his attentions is left unclear. A few years after the Crests, Neil Sedaka delivered the archetypal "Happy Birthday Sweet Sixteen," a smooth scoop of vanilla dotted with marshmallow tra-la-las that's more like a father's tribute to a daughter than a declaration of romantic love.

> "Any halfway decent girl can rob me blind, because I'm too torqued up to say no."
> —The Geek (Anthony Michael Hall), *Sixteen Candles*

Channeling the torqued-up fever dreams of geekish adolescents is a rock tradition. Johnny Burnette's "You're Sixteen" expertly conveys this full-on, weak-in-the-knees infatuation/lust hybrid. Burnette's rockabilly roots ran somewhat counter to this shiny, over-perky concoction, but the song remains his signature. A former schoolmate of Elvis Presley, and his contemporary in the local music scene, Burnette was reportedly turned down by Sam Phillips of Sun Records for sounding too much like Sun's new hire.[2] Ringo Starr's (looser, kazoo-ier) remake of "You're Sixteen" in 1974 gave Mr. Starkey his second consecutive No. 1 single (after "Photograph").

Sam Cooke sings "Only 16" in a voice so pure that it seems he's trying too hard to be "good," as if letting out a single lustful melisma would give the game away. The song's most memorable line—"She was too young to fall in love/and I was too young to know"—contains a novel-length narrative in 15 words. Dr. Hook & the Medicine Show gave the song a warm soft-rock bath in 1976 and nearly topped the charts. Oddly enough, it took the Persuasions, a venerable a cappella group who've covered the spiritual canon as well as the Dead, Zappa, and U2, to grab hold of "Only 16" and really let it rip.

---

[1] B.B. King's "Sweet Sixteen," is an exception ("Treat me mean, baby/But I'll keep loving you just the same"). So is "A Young Girl (of Sixteen)," written by Charles Aznavour and sung by Noel Harrison (son of Rex) in 1966, whose subject dies for the love of a cad: "A child of springtime, still green, lying here by the side of the road."

[2] Elvis, coincidentally, died on August 16, 1977 and sang memorably about a mystery train "16 coaches long."

David Klein

The air of innocence surrounding Sweet 16 has dissipated in recent decades. By the time Fall Out Boy recorded "A Little Less Sixteen Candles, a Little More 'Touch Me,'" the 16 vestal virgins from Procol Harum's "A Whiter Shade of Pale" had long since left for the coast. In No Doubt's "Sixteen," Gwen Stefani toys with a teen boy who "can't cop a feel." The protagonist in Le Tigre's "Sixteen" is wise to love's self-delusion but still willing to fall into it.

Back in the '70s, though, Sweet 16 had no such self-awareness or empowerment. In fact, the term was used almost solely on a salacious basis. The cover of *16 and Savaged*, by second-tier British glam rockers Silverhead, depicts a tarty teenage girl whose face (and cleavage) is reflected in the mirror of a pub restroom.³ In the fertile Berlin spring of 1977, Iggy Pop recorded "Sixteen," his lone composition on the Bowie-helmed *Lust For Life* and a lowdown cry from the loins: "Show you my explosion, sweet sixteen!!"

Meanwhile, across the pond at New York's Record Plant recording studio, the members of Kiss were laying down tracks for "Christine Sixteen." The original demo featured a solo by a young hotshot named Eddie Van Halen, which Gene Simmons insisted that Ace Frehley re-create in full. Making the most of a sweetly stinging, Stones-y guitar lick, "Christine Sixteen" is my favorite song by Kiss, a band whose musical appeal I usually can't fathom. (For some reason, I picture a listless suburban mom and her son's driver's-ed teacher cranking this track on the way to their motel tryst near the Lincoln Tunnel, like a scene from a lost Todd Solondz movie.) In any case, "Christine Sixteen" simply kicks ass. Maybe it's that saloon piano going clang-clang-clang-clang. Or those three snare drum hits that set off the chorus. Even Paul Stanley's lunkheaded soliloquy in the middle makes me want to recite right along with it. Try it sometime.

The Replacements covered Kiss's "Black Diamond" on their 1984 masterpiece, *Let It Be*, when most in the indie world dismissed bands like Kiss out of hand. But even as Paul Westerberg remained steadfast in his admiration for Kiss, his musical palette continued to deepen, despite the protestations of rhythm guitarist Bob Stinson, who would advise him to "save that for your solo record." It does not seem like a leap of logic to assume that "Sixteen Blue" was one of the first Westerberg songs to really piss off Bob Stinson.⁴ All traces of hardcore, or Thin Lizzy for that matter, are long gone, replaced by a laid-back country tempo, Westerberg's raw, vulnerable vocals, and a message of empathy decked out in shimmering Byrdsian spangles. (An enterprising Warner Music A&R man reportedly offered the song to Rod Stewart.)

Sixteen is no mere bastion of raging hormones. "Sixteen Tons," the Merle Travis song that became a country hit in 1946 and which Tennessee Ernie Ford took to No. 1 in 1955, has nothing to do with cakes, candles, unrequited love or underage sex.

MATCH THE #16 LYRIC TO ITS SOURCE

1) "I said hallelujah to the sixteen loyal fans"	A) Morrissey, "Late Night Maudlin Street"
2) "Sixteen stitches put her right and her dad said 'don't say I didn't warn ya'"	B) Ian Hunter, "Irene Wilde"
3) "When I was just sixteen/ I stood waiting for a queen"	C) The Who, "Sally Simpson"
4) "No one stays up for you when you have sixteen stitches all around your head"	D) The Sixths, "All Dressed Up in Dreams"
5) "So I cried for 16 days/ and I cried in 16 ways"	E) Lo-Fidelity All-Stars, "Battleflag"

Answers: 1-E, 2-C, 3-B, 4-A, 5-D

---

3   Siouxsie & the Banshees' "Circle" takes an unflinching look at early promiscuity, as a "pretty girl of 16" becomes a "ruined girl of 16" due to pregnancy.
4   It's true that Westerberg had included the glorious ballad "Within Your Reach" on the band's previous outing, *Hootenanny*, but he played all the instruments himself.

It's a mining song-cum-biblical parable, the story of a man who cannot escape his fate. Fittingly enough, Johnny Cash sang "Sixteen Tons," as did Bo Diddley, Stevie Wonder, and at least 16 more (including Lorne Greene). The song all but cries out for a manly delivery, and few women have signed on for the job.[5]

And what a strange standard it is. Muscle and blood, skin and bones, a man who can't even die because he owes too much money—all par for the course on the country charts, but far from your typical mainstream fodder. Ford's version triumphs through its spare arrangement, which provides the bedrock for the singer's subterranean baritone; rich, resonant and textured—some have called it booming. When he winds up with a vocal flourish on the last "I owe my soul," for a second or two we are pulled along solely by that voice, momentarily and powerfully a cappella, a rarity for a commercial radio single.

In "June 16," the Minutemen pay homage to James Joyce's *Ulysses* in their preferred form: 1:49 of astringent funk. For funk of a more backbone-loosening variety, look no further than "The Funky Sixteen Corners" by the Highlighters Band, a dance lesson/party on wheels led by vocalist James Bell, who exhorts, howls, and teaches in the manner perfected by another man with the initials J.B. As for those 16 corners: "That's funky four to your left, funky four to the right, funky four to the back, and funky four to the front."

CONNIE BE GOOD: "Sixteen Reasons," a laundry-lister that Connie Stevens took to No. 3 in 1960, has enjoyed an unexpected second life thanks to a bewitching scene in David Lynch's *Mulholland Drive*. The husband-and-wife duo of Bill and Doree Post wrote and performed the original, but it was the version by the former Concetta Rosalie Ann Ingoglia that lit up the charts. Still, hats off to the Posts. By listing all 16 reasons (including "the way you comb your hair" and "the crazy clothes you wear") they schooled the likes of Chuck "Not Quite 13 Question Method" Berry and Paul "Nowhere Near 50 Ways to Leave Your Lover" Simon.

> "Like John Lennon says, 'Anything that's chonka, chonka, chonka, chonk is Chuck Berry.'"
> —Chuck Berry, *Rolling Stone* interview with Paul William Salvo, 1972

Chonka chonka chonka chonk proves the point that writing about music is like dancing about architecture. Words often fall short, but chonka chonk endures. Like any of Berry's immortal early classics, "Sweet Little Sixteen" has this sound, along with a touch of Latin influence that he said resulted from "the mambo-calypso thing [that] was going around." If you don't hear it, try repeating "cha-cha-cha" along with the song and it will become clearer. "Sweet Little Sixteen" shows why Berry is not only the original rock guitar hero but also one of the premier lyricists. The travelogue motif he uses here, and would return to in "Living in the U.S.A.," has become part of rock's vocabulary, from the Beach Boys' "Surfin' U.S.A." to Steve Miller's "Jet Airliner" ("Philadelphia-Atlanta-L.A.") and onward.[6]

Berry weaves his litany of cities around a sparse narrative: A girl spends all her waking hours obsessing about her favorite rock 'n' roll groups while slogging through her humdrum teenager-hood. Like a novelist, Berry doesn't tell us, he shows us what it means to love rock 'n' roll, as well as the place it occupied in the hearts of America's youth in the late '50s. He also illuminates the

---

5 Except Anna Domino, whose real bid for number-song glory lies ahead.
6 Not that Berry invented the travelogue; the prototypical "Route 66," for example, was a hit for Nat King Cole in 1946, the same year, coincidentally, that Merle Travis wrote "Sixteen Tons."

complexity of the lives of young people. The fan at the heart of "Sweet Little Sixteen," after all, has "the grown-up blues," not just in the sense of tight dresses and lipstick but in her bifurcated existence; by night she's a full-fledged music obsessive, by day she's just another high school kid shouldering society's demands along with those of her parents.

THE VERDICT

Forced to choose between "Sweet Little Sixteen" and "Sixteen Blue," I have to go with Chuck. The reasoning is simple: no Chuck Berry, no Replacements. This song is primal. Even "Mommy Mommy" and "Daddy Daddy" make appearances. In his memoirs, Keith Richards fondly recalls Berry's performance of "Sweet Little Sixteen" in *Jazz on a Summer's Day*, a film, he writes, that "drove a thousand musicians." And Berry didn't just predate the 'Mats; he was a strong influence. The Replacements are Chuck's children[7] as much as Chilton's. In sum, the ever-quotable John Lennon ventured beyond chonka chonk when he paid Berry the ultimate compliment: "If you tried to give rock 'n' roll another name, you might call it Chuck Berry."

A SONG TITLE	AN ALBUM	A BAND	A LYRIC
"Sixteen" – The Buzzcocks (1978)	*16 Lovers Lane* – The Go-Betweens (1988)	16 Horsepower – Denver, Colo., alt-country (1990s–2010s)	"They do all 16 dances" – The B-52's, "Dance This Mess Around" (1979)

---

7 The phrase "Chuck's children" belongs to Bob Seger, as in "all of Chuck's children are out their playin' his licks." The source of the lyric, "Rock and Roll Never Forgets," also contains the line "Sweet sixteen's turned 31." In his compendium *Rock and Roll Always Forgets*, Chuck Eddy extols John Cougar Mellencamp's riff on the aforementioned Segerism: "Seventeen has turned thirty-five."

"I am sixteen going on seventeen/ I know that I'm naïve"
—"Sixteen Going on Seventeen," *The Sound of Music*

By the time Sweet Little Sixteen reaches the Edge of Seventeen, to hear many a songwriter tell it, she's done a lot of living. Perhaps she's even gotten herself hitched, like the young bride-to-be in Marty Robbins's sober benediction "She Was Seventeen (He Was One Year More)." Still, most #17 songs deal with more typically teen-centric issues. The mother of them all might well be "At Seventeen" by Janis Ian, who as a teenager recorded "Society's Child," an interracial romance-themed song and an unlikely hit in 1967. Nearly a decade later, Ian brought adolescent angst to the upper reaches of the charts in all its squirmy glory: "And those whose names were never called/When choosing sides for basketball" still has the power to stir up squalls of douche-chills among both the familiar and the uninitiated.

More than a few writers have taken special pleasure in despising "At Seventeen," which began a lengthy radio run in early September 1976—just in time for the new school year. It holds an esteemed spot in both Guterman & O'Donnell's *The Worst Rock 'n' Roll Records of All Time* ("Maybe if you stopped whining about valentines and started concentrating on your perimeter shot, you'd have a better chance") and *I Hate Myself and I Want to Die: The 52 Most Depressing Songs You've Ever Heard*, in which author Tom Reynolds points out that "At Seventeen" is "the only depressing song of note based on a samba beat." In fairness, the song has occasionally been treated with dignity, such as by Jeffrey Eugenides in *The Virgin Suicides* and in the film *Mean Girls*. But as long as the likes of Celine Dion are moved to cover it, the parodies will continue.

Because of its ubiquity on the airwaves during the singer-songwriter-friendly '70s, "At Seventeen" was pretty much the only #17 song that mattered for a good year or two. But Ms. Ian was ultimately sent scurrying back to the library by songs like the Sex Pistols' "Seventeen," one of the less celebrated sound bombs from *Never Mind the Bollocks, Here's the Sex Pistols*. Structurally it's closely related to "God Save the Queen," without reaching the incendiary heights of that blazing rocker. On the chorus, Rotten-Lydon bellows, "I'm a lazy sod!"—a nod to the lack of industry preferred by many 17-year-olds—even though the song's object of scorn is a hoary 29.

That's still a relatively upbeat scenario next to the Cure's "Seventeen Seconds," named for the final moments of life before a self-induced death. The sound is minimal, just ice-water guitar chords and a skeletal drumbeat, but gradually it gets, you know, catchy. Under a YouTube clip of the song, "dolceluongo" summed it up best: "a great song … to kill yourself."

Maybe the Stray Cats were sick of all this negativity when they came out with "(She's) Sexy + 17," reminding the world that, hey, teenage girls can be awfully attractive. Rick James basically agreed, in his own song called "17," although, to get technical, the saucy Stray Cats gal is sexy *and* 17 while Rick's is "seventeen *but* she was sexy."

The joyous couplet that begins the Beatles' "I Saw Her Standing There" ("Well she was just seventeen/you know what I mean")—along with "I can't get no satisfaction" and a few others—surely ranks among the great opening lines in rock. There's a world of innuendo packed into that "you know what I mean."[1] Paul McCartney had initially written a far tamer second line ("Never been a beauty queen") but changed it when his writing partner called it corny. Maybe the phrase (and its working title, "Seventeen") reminded Lennon too much of a lightweight 1955 hit, also called "Seventeen," by a rockabilly purveyor named Boyd Bennett: "Seventeen, hot rod queen/Cutest girl you've ever seen." Had Lennon and McCartney opted to use the seventh prime in the title, they would have been in good company. Why, if a group of artists who've recorded songs called "Seventeen" all linked arms and danced around the maypole, that ring would include Bobby Brown, the Troggs, Jethro Tull, Winger, Tim McGraw, the Eurythmics, and the Kings of Leon. And oh, what a sight it would be.

Songwriters of all stripes turn to 17 to fulfill an inflection that 16 or 18 can't supply. Tracey Ullman's "You Broke My Heart in 17 Places," for example, whose chorus adds "Shepherd's Bush was only one" to the title phrase, could not exist without those three beats. Neither could Meat Loaf's "We were barely seventeen and we were barely dressed," or the Cars' "And she won't give up/'cause she's seventeen," both of which stoked teen dreams in the late 1970s. Fountains of Wayne's "Joe Rey" has "seventeen different words for snow"; the joy riders in Marshall Crenshaw's "Rockin' Around in NYC" cruise through "seventeen lights in a row." And in "My Bag," Lloyd Cole proclaims, regarding him and his good thing: "We gave up sleep at the age of seventeen."

Rising from some strange, pillowy planet is Broken Social Scene's "Anthems For a Seventeen-Year-Old Girl," sung by Emily Haines (also of Metric), whose heavily processed voice calls to mind a robot geisha. The singsong melody hops between two notes while a subtly rippling banjo and the long tones of a sympathetic string quartet act as connective tissue. There are no drums. "Park that car, drop that phone, sleep on the floor, dream about me," she entreats, as if making up a song while playing cat's cradle and watching it snow out the window.

In an altogether chillier enclave is Liverpool's Ladytron, makers of gelid electronic dance music. In "Seventeen," a viscerally sexy beat undergirds the blasé metallic whispers of Helen Marnie and Mira Aroyo: "They only want you when you're seventeen … When you're twenty-one you're no fun." (Charlotte Gainsbourg's "The Songs That We Sing" takes it even further: "I read a magazine/that said by seventeen/ your life was at an end/I'm dead and I'm perfectly content.") Keren Ann, the singer-songwriter born Keren Ann Zeidel, channels Leonard Cohen's sophisticated world-weariness in the stately "Seventeen" and achieves sublimity. But her impossible precociousness belies the song's central plaint: "Look at me/I'm only seventeen." This is an *exceptional* 17, the smartest girl in the class, editor of the poetry mag and not at all typical.

THE VERDICT

The definitive #17 song has to embody the whole passionate, confused, on the cusp of something significant but not quite there 17-ian ethos. That's why there's only one real choice for me. Announcing itself boldly in a spray of 16th notes, Stevie Nicks' "Edge

---

[1] Jeff Beck clearly found the phrase inspirational. His watershed *Blow By Blow*, produced by George Martin himself, opens outlandishly well with "You Know What I Mean" before seguing into a cover of the Beatles' "She's a Woman."

of Seventeen" has the feverish intensity of a confused, hormonal, diaphanously clad teenager. But it's heavier than that. Written in 1981, the song draws from Ms. Nicks's dual sense of grief over the death of John Lennon and the deteriorating condition of her beloved uncle. The title came first. Tom Petty's wife referred with a Southern inflection to "the age of seventeen," and to Stevie's ears "age" sounded like "edge." A monster was born.

Somehow it's appropriate that a song based on a mishearing (see also the previously mentioned "Seven Nation Army" and "In-A-Gadda-Da-Vida") revolves around a refrain that is itself often misheard. Especially since the mishearing makes sense, unlike say, hearing the Go-Go's "Our Lips Are Sealed" as "Alex the Seal." For what is a practically 17-year-old but a one-winged dove that cannot fly—or rather *can* fly, but only round and round without getting anywhere? But nope—it's a *white-winged* dove[2] that sings a song sounds like she's singin' *ooh, ooh, ooh*.

According to Waddy Wachtel, who played guitar on "Edge of Seventeen" and the rest of *Belladonna*, the Police's "Bring on the Night" was the primary inspiration for the feel of "Edge." Beyond inspiration, in fact; Wachtel has made no bones about his belief that he and his fellow musicians crossed the line into straight-up musical theft. Listening to the two songs now, the similarities cannot be denied, yet each has a distinct aura. Sting's a Buddhist anyway; surely he's forgiven Ms. Nicks for appropriating a riff or three.

A SONG TITLE	AN ALBUM	A BAND	A LYRIC
"Opus 17" – The Four Seasons (1966)	*Seventeen Seconds* – The Cure (1980)	Heaven 17 – Sheffield, England, synth-pop (1970s–1980s)	"In some heaven whose number is seventeen" – Be Bop Deluxe, "Love Is Swift Arrows" (1974)

---

2 *Zenaida asiatica* (purportedly the working title of the Police's *Ghost in the Machine*)

18

Even a casual, non-obsessive music listener could come up with the definitive #18 song without too much contemplation. "I'm Eighteen" by Alice Cooper is the obvious choice, and it looms extra large in the absence of strong contenders. Oh, there are plenty of #18 songs out there—just not a ton of great ones. When big names do show up, they don't bring their A-list material. The gently lilting "18 Yellow Roses" finds Bobby Darin sounding uncannily like Roy Orbison, but why? Prince[1] could write a good song in his sleep, and "18 and Over" (rhymes with "I wants to bone ya") doesn't disprove that. Originally slotted for *Sign O' the Times*, which Prince envisioned as a triple album but whittled down, under protest, to a double, the song eventually appeared on *Crystal Ball*, a whopping 42-song collection released in 1998. Still, one would be hard-pressed to call it essential Prince. Similarly, Cat Stevens' "18th Avenue (Kansas City Nightmare)," a suite-like chamber rocker from the chart-topping *Catch Bull at Four*, finds Stevens in familiar gruff-voiced form but falls short of essential Cat. On the other hand, one could make a strong argument that "Hangar 18" is essential Megadeth.

"18th Street Strut" is a jaunty stride piano tune recorded by Fats Waller in 1926. The title of Rakim's "The 18th Letter (Always and Forever)" is an alphabetical reference to the renowned emcee's stage name. In need of rediscovery is "18 Carat Love Affair" by the Associates, a U.K. hit in 1982 for this short-lived Scottish synth-pop outfit led by the mercurial, ultimately tragic singer Billy McKenzie.

"18 With a Bullet" was an unlikely hit for English session man Pete Wingfield—it actually reached #15 on the U.S. pop chart in 1975, though not necessarily with a bullet. Sung mainly in falsetto, this doo-wop-style ditty turns record industry-speak into love talk. "I'm a national breakout," Wingfield warbles, presaging *Midnite Vultures*-era Beck by about 25 years, "so let me check your playlist, mama/ Come on, let's make out." It was Wingfield's lone solo charting, but he went on to produce Dexy's Midnight Runners and the Proclaimers, and was the man behind Mel Brooks's "It's Good to Be the King Rap (Part I)."[2]

Eighteen-wheelers, which have been so named since the mid-'70s, have inspired at least 18 songs. "18 Wheels of Love" by Drive-By Truckers opens with, "Mama ran off with a trucker" and ends with a Dollywood wedding officiated by a Porter Wagoner look-alike. And unlike heart-tugging No. 1 hits for Kathy Mattea ("Eighteen Wheels and a Dozen Roses") and Alabama's "Roll On (Eighteen Wheeler)," it's completely autobiographical.

SHLOCK ROCK TRIFECTA
VERSION #18
Bryan Adams – "18 Till I Die"
Skid Row – "18 and Life"
Black 'N Blue – "Hold On to 18"

---

1   Boy George once likened him to "a dwarf who's been dipped in a bucket of pubic hair," but who got the last laugh?
2   Wingfield should not be confused with Whigfield, the nom de pop of Sannie Charlotte Clarkson, a Danish vocalist of modest gifts who scored an unlikely U.K. No. 1 with "Saturday Night" in 1994.

Pink, whose parents had to sign her first record contract because she was under 18, runs romantic notions off the road with "18 Wheeler," her sassy ode to resilience from *Missundaztood*.

NEITHER FISH NOR FOWL: ? & the Mysterians' "'8'Teen," a 1966 B-side, can neither be called a true #18 song nor a true #8 song. But numerical definitiveness, not to mention glory, awaits at 96.

THE VERDICT

We take provocation as a given now, but in the era in which Alice Cooper came to prominence (with a little help from Frank Zappa), the scruffy Rolling Stones were considered the bad boys of rock, and they never killed a single chicken. "I never really did anything that outrageous on stage," the former Vincent Furnier told Lester Bangs in 1975. "It's just the fact that it had rock 'n' roll behind it that made it sound so damn notorious."

Sure, but all spectacle aside, "I'm Eighteen"—on the strength of Alice's take-no-prisoners vocal, that central serpentine guitar line, the keening harmonica, and the head of steam kicked up by the band—is the kind of lean, ageless rock single that confused kids can still snarl along with. Many sources erroneously indicate that Cooper's "Eighteen" is based on "'8' Teen" by ? & the Mysterians. To be sure, Cooper, who wrote the lyrics when he was a full 23, could easily have heard the song by his fellow Michiganders while living in Detroit, but musically there's no comparison. For singer Rudy Martinez (who had his name legally changed to ?), 18 is just a lyrical launching pad ("I had 18 arms to hold you/18 lips to kiss you/18 ways to love you"). Meanwhile, Alice is *eighteen*. Big difference.

Blunt musical expressions of teenage anger and frustration are rock 'n' roll's common currency, but few are so palpably delighted with themselves as "I'm Eighteen." No wonder it was the song that the young, soon-to-be-Rotten John Lydon chose to lip-synch to when he auditioned for the Sex Pistols. Even in Britain's laddish youth culture, Alice Cooper, a mascara-wearing American with a girls' name, was widely admired, even imitated. As Lydon reported in his first published memoir, *Rotten: No Irish, No Blacks, No Dogs*, young fans of the Newcastle football club liked to dress like Alice, eye makeup and all.

A SONG TITLE	AN ALBUM	A BAND	A LYRIC
"18" – Moby (2002)	*Music for 18 Musicians* – Steve Reich (1978)	18th Dye – Germany/ Denmark, noise rock (1990s)	"He was just 18/ proud and brave" – The Band, "The Night They Drove Old Dixie Down" (1969)

LAST WORDS:
"I've had eighteen straight whiskies. I think that's the record …"—Dylan Thomas, 1953, actual last words

19

Nineteen raises what's known in the rock numerology trade as the Prince Conundrum. In other words, is "1999" eligible? It does have a 19 in it, but not a stand-alone 19. One thousand nine hundred and ninety-nine is what it really is. Guided By Voices get around this with "Dayton, Ohio – 19 Something and 5," but the average year-titled tune cannot be considered a true #19 song. We're dealing with numbers here, and if we flout the specificity of each one, we abandon the unshakable sense of order they provide in the first place. Should someone ever decide to chart number songs up into the high thousands, John Cale will perch regally atop "Paris 1919," just south of the Who, who will rule "1921." You might hear Harry Nilsson singing "1941," and Richard Thompson (at "1952") revving up his Vincent Black Lightning. New Order at "1963" will send semaphores to Robyn Hitchcock at "1974" (ignoring Iggy Pop's frenzied "boo-hoos" in "1969"), whose smoke is bothering Smashing Pumpkins five slots up at "1979." The members of Spirit and David Bowie will be wrestling over rights to "1984," Luna will be chilling at "1995," and on it will go, to the year 2525 and beyond even R.E.M.'s "1,000,000." But right now, let's stick with the proper 19s. It's not like there aren't any.

On "Hey Nineteen," all Steely Dan's obvious gifts are on display, only more subtly than in much of their oeuvre.[1] There are tasty licks aplenty here, but also a sense of space and an absence of fancy-pants soloing. The song's loose gait fits its aging former frat boy narrator like a well-tailored suit. He's in crisis because his wide-eyed date has never heard of Aretha Franklin, but that's nothing a little weed and tequila can't fix. The sweet moment comes as the band rides the groove for an extra few bars right before the vocal cream ("the Cuervo Gold") is poured. Make no mistake though: those bars were plotted out weeks in advance in a secret bunker.

"Hey Nineteen" is a fascinating period piece. Colombian herb is no longer prized, nor is Cuervo considered high-end stuff anymore. And apparently, in 1980 a major music industry act like Steely Dan could make the Top 10 with a single that both proselytizes on behalf of pot 'n' booze and ranks on uninformed nubiles, with exactly zero controversy. How times have changed.

There's a world of difference between "only 19" and "just 19." Only-19 songs, like the pair of ditties called "Nineteen" by the Old 97s and Buck O Nine, refer to 19 in the naïve, "just finished high school" sense. In the other camp is the Eagles of Death Metal's "I Got a Feeling (Just Nineteen)," a pre-coital victory dance built around a couple of sludgy chords, the campy falsetto vocals of Jesse "The Devil" Hughes, and the band's signature "stripper drum beats." Here, "just nineteen" means "just right."

When you hear a phrase like "regularly cited as one of the greatest pure rock stars of all time" or "Along with Bob Dylan, Jim Morrison, Patti Smith and Bruce Springsteen … one of the first to merge poetry with rock music," the name Phil Lynott might not spring to mind. Yet there it is on AllMusic.com, and it's worth taking a deeper look at the complicated lead singer of '70s twin-guitar gods Thin Lizzy. By the early '80s, when the band was deep into its death spiral, Lynott put out a pair of

---

[1] You really shouldn't ever use the word "oeuvre" *except* when referring to Steely Dan.

solo records with Dire Straits' Mark Knopfler as a sideman. A few years later, the solo bid scrapped, Lynott found a new band, Grand Slam. "Nineteen" is all chest-pounding proclamations, motorcycle revs, and artless shredding—certainly not an apt final statement for Lynott, who died not long thereafter at 36 (not even twice 19) from a lifetime of excesses.

In an uncanny instance of numerological serendipity, the producer of "Nineteen," Paul Hardcastle, had an international hit earlier that year with, you guessed it, "19." Built around words from a TV documentary about war veterans and incorporating an audacious use of then-novel sampling technology, "19" constituted a serious attempt to raise consciousness about post-traumatic stress disorder. And you could dance to it.[2]

THE VERDICT

Accusations of witchcraft are about the only charge Mick Jagger doesn't hurl at the hapless society girl he eviscerates in "19th Nervous Breakdown," as mean-hearted as any in the rich trove of misogynistic Stones songs. Written about the same person who inspired such musical daggers as "Stupid Girl" and "Under My Thumb" (Chrissie Shrimpton, sister of the model Jean), this is a harsh takedown of a woman who could only be English, by a man who could only be English. Indeed, during a brief but fertile few years, Rolling Stones songs were suffused with an unmistakably British flavor. By the end of the decade, that locale had been replaced by the "ballrooms and smelly bordellos/and dressing rooms filled with parasites" of *Exile on Main Street*. But for a few years, before things got torn and frayed, it was all Lady Jane and St. John's Wood, windscreens and Union Jacks.

"19th Nervous Breakdown" comes from a particularly fertile streak of singles produced by Andrew Loog Oldham in the mid-'60s, as the Stones made their transition from scruffy American blues obsessives/copyists into the self-proclaimed "world's greatest rock 'n' roll band." The number 19 means nothing here, and it doesn't have to. "I thought of the title first—it just sounded good," said Mick Jagger at the time, and it makes for a great bit of alliteration in the chorus of a song that's truly worthy of the label "classic Stones." And what a slice of sheer perfection it is, from that alchemical explosion at the very beginning, when one guitar line slices in for two bars before being joined by a second phrase, like an electron orbiting a nucleus. Bass and drums kick in, then Charlie Watts's jittery cymbals, making a perfect launching pad for a searing Jagger vocal. There's that jarring Keef corkscrew guitar lick right before "Here it comes," the marvelous middle eight ("Oh, who's to blame?"), and Bill Wyman's TNT-dropping bass runs in the fadeout.

Character assassination has never sounded this good.

A SONG TITLE	AN ALBUM	A BAND	A LYRIC
"19 & Mad" – Leyton Buzzards (1978)	*19* – Adele (2008)	The Nineteenth Whole – Indianapolis, soul-jazz (1970s)	"There's one for you, nineteen for me" – The Beatles, "Taxman" (1966)

---

[2] It also boasts one of the most unlikely opening lines ever to grace a hit song ("In 1965, Vietnam seemed like just another foreign war, but it wasn't"). Not quite "Well, since my baby left me/ well, I found a new place to dwell," is it?

20

Before the century was half over, Noël Coward had seen enough of it to write "20th Century Blues." Elton John, Marianne Faithfull, Charlie Parker, and Robin Tower have all covered Coward's lament for the "hurly-burly of insanity" into which he felt his epoch had devolved. In 1971, Ray Davies of the Kinks expressed similar sentiments in "20th Century Man," echoing Coward with "the age of insanity" while upping the ante with napalm and power-mad civil servants. Of course, Ray had the advantage of hindsight, which is always 20/20.[1]

The Coward connection shouldn't come as too much of a surprise. One of Davies' several vocal tonalities—his crooning, slightly dissolute sophisto voice—has a distinctly Cowardian flavor, and many a Davies lyric ("custard-pie appreciation consortium," for example) possesses a similar acuity for English life, while requiring a level of elocutionary skill worthy of Sir Noël. Certainly Davies' withering eye for the foibles of mankind, on display in such songs as "Dedicated Follower of Fashion," isn't so far from the drolleries of the man who wrote "Don't Put Your Daughter on the Stage, Mrs. Worthington."

In an opposite corner of the universe prowls the Doors' "Twentieth Century Fox." It's no "Light My Fire," but this is far more satisfying than other latter-era expressions of Lizard King lust, like "Touch Me" and its bombastic Vegas horns. Clocking in at a mere 2:30, this is a model of concision for a band whose songs grew longer in tandem with the lead singer's beard. One can quibble with the logic of a "fashionably lean, fashionably late" fox, who has the world "locked up inside a plastic box" yet will never break a date—when that is *exactly* what foxes do. But it swings, and "No tears, no fears/no ruined years, no clocks" is a damn good line. The fox in the .38 Special song of the same name is just a spoiled brat in designer jeans and a powder blue Corvette, but the Doors and the Spesh do agree on one thing—the proper way of appreciating a 20th-century fox: "Just watch the way she walks."

Jim Morrison and Marc Bolan had much in common: rock stardom, androgynous good looks, early deaths, and a "20th-century" song. I've never seen a T. Rex best-of that didn't include "20th Century Boy," and for good reason: it's one of a handful of great singles by one of the primo singles bands of the '70s. "20th Century Boy"'s Cro-Magnon central hook and the wallop of Tony Visconti's production would sound right in any era, which explains why a decade-spanning array of singers, from Siouxsie Sioux to the Replacements to Adam Lambert, have been keen to lay into this 1973 pile-driver. It's been canonized in a Levi's ad, in Guitar Hero, and on film (the fictitious Flaming Creatures cover it in *Velvet Goldmine*). Bolan even had a solo song called "20th Century Baby" among the avalanche of unissued tracks he left behind, but it does not approach the "Boy."

The story goes that James Taylor's "Suite For 20 G" was a contract-fulfilling creation composed of various musical fragments stitched together into a song for the sole purpose of recouping $20,000 from his record company. And Taylor, as savvy in music

---

[1] Morrissey, another acerbic Englishman, wrote in 1986, "I feel the 21st century breathing down my neck."

as in business, gets away with it. The Beatles originated this type of patchwork construction on side 2 of *Abbey Road*, although the transformation of musical odds and ends into lengthy suite-like compositions certainly qualifies as one of the Fab Four's less essential legacies.

A pre-Quarrymen Paul McCartney performed Eddie Cochran's "Twenty Flight Rock" for his future songwriting partner in order to impress him, because the song requires that a player be fleet both of tongue and fingers. The Minnesota-born Cochran was blessed with musical talent and the smoldering looks of a pompadoured early rocker. He was a successful session guitarist who'd given the world two classic singles in "Summertime Blues" and "C'mon Everybody" at the time of his death in a London taxicab at 21. Like Buddy Holly and Ritchie Valens, Cochran was a real teenager, writing songs for and about teenagers.

THE VERDICT

"Twenty Flight Rock," though less well-known than Cochran's signature hits, is a damn fine expression of the late rockabilly form, one that the Stones and Jeff Beck and others have seen fit to cover. And, as a title, "Twenty Flight Rock" transcends its rather mundane attachment to the floors in Eddie's apartment. It could almost be a reference to the level of octane contained in the song. Like, "This here's some 20-flight rock. I'd take a step back if I was you."

T. Rex's "20th Century Boy" is a colossus that captures Bolan's pandrogynous, shape-shifting appeal and leverages the number's key association—the century thing—for all it's worth. Now that we have entered the Twenty Zone, it seems right to mark the occasion by honoring a song from the second wave (or third, depending on how you slice it) of rockers. In reconfiguring rock during its early Me Decade adolescence, Marc Bolan and T. Rex drew liberally from styles dating back to Cochran's era and, with the addition of eyeliner and Les Pauls, injected a needed dose of uncomplicated fun into a medium that had already begun taking itself way too seriously.

A SONG TITLE	AN ALBUM	A BAND	A LYRIC
"Twenty" – The Undertones (1979)	*20 Jazz Funk Greats* – Throbbing Gristle (1979)	Matchbox Twenty – Orlando, Fla., alt-rock (1990s–2010s)	"They sentenced me to twenty years of boredom" – Leonard Cohen, "First We Take Manhattan" (1988)

21

"How could I possibly know what I want when I was only 21?" wondered Sinead O'Connor in "The Emperor's New Clothes." At 21, a life may hold infinite promise ("She was 21/when I left Galveston"[1]) or it may already be dramatically circumscribed ("I turned 21 in prison doing life without parole"[2]) To most First Worlders, the age of 21 signifies full-fledged adulthood. In the words of Justin Bieber, "21's big because you can legally go out and, like, party and stuff."

In "Three Times Seven," Merle Travis expressed the essence of being 21, 1959-style: "I'm three times seven/and I do as I doggone please." Fittingly enough, it was the Eagles who stepped up and addressed what 21 felt like in the early 1970s. Written and sung by Bernie Leadon, the aggrieved founding Eagle who wanted to keep it real, "Twenty-One" typifies the empty sloganeering of the rudderless early '70s ("21 and strong as I can be/I know what freedom means to me"). Still, Leadon's chosen method of leaving the nest has stood the test of time. Disgusted with the nakedly commercial direction the band had taken, Leadon made his departure official by pouring a beer over Glen Frey's talented head.

Well before we reached the 21st century, the era had captured the imagination of numerous songwriters. Roughly 15 years in, no song seems to have captured Century #21 definitively, although the Red Hot Chili Peppers and Bad Religion are among those who have tried. King Crimson's caustic "21st Century Schizoid Man" is at the head of the pack. Featuring the treated vocals of Greg Lake savagely spitting out disturbing lyrics penned by Pete Sinfield (producer of the first Roxy Music record), it sets the stage for *In the Court of the Crimson King,* one of prog rock's undisputed triumphs. Kanye West, a man of refined taste in samples, incorporates the song's title phrase in "POWER" from *My Beautiful Dark Twisted Fantasy.*

Chuck Berry's "21" comes from his period of greatest innovation but only came to light fairly recently, on a release of Berry's complete 1950s Chess recordings. It has a carefree quality that perfectly captures the mind-set of its young narrator. The recording hasn't suffered from excessive "improvement" either; when the song opens with one of Chuck's patented intro licks, it's accompanied by a refreshing overtone of distortion, and the snare drum sounds like a garbage can lid.

The music is buoyant enough, with a rhythmic scheme similar to "Maybelline," but the sentiment is rather tame for a Chuck Berry song. The singer is a practical guy who can tamp down his lust and wait patiently for his sweetheart to turn 21 so he can marry her. "So we'll dance and bypass romance," he sings, "right now they say we're too young." In the meantime, this uncharacteristically obedient Berry stand-in is content to partake in the joys of picnicking, swimming, and other forms of chaste courtship. It gets the foot tapping but strains credulity.

---

1 "Galveston" by Jimmy Webb
2 "Mama Tried" by Merle Haggard

If 6 Was 9

THE VERDICT

While the suitor in Berry's "21" is willing to wait for society's blessing, the young woman belting out the Shirelles' "Twenty-One" has no time for delayed gratification. The song's caffeinated clip and string-heavy arrangement give it a touch of camp, and, true to its era, the final verse tries to rein in the singer's ardor (she imagines one day telling her own kids that they, too, must wait until they reach 21 to "paint the town red"). But "Twenty-One" is still plenty sassy and serves as a fine showcase for the giddy joys of this proto-girl group.

The Shirelles, it should be said, were four young women who had hits with songs by the young Burt Bacharach and the stellar team of Gerry Goffin and Carole King, among others, in the early '60s. They were favorites of the Beatles, George Harrison especially. On *Please Please Me*, the Beatles covered a pair of songs made popular by the Shirelles. Of course, it was the seismic shift brought on by those very same Beatles that made the musical landscape so inhospitable for the Shirelles. "Twenty-One," a B-side written by Luther Dixon, who also penned "Sixteen Candles," "Mama Said," and "Soldier Boy," is a credible slice of teenage life circa 1961. Lead singer Shirley Alston sounds girlish yet self-possessed here (dig that tossed-off "steady now" before the fade). She is not looking for approval. She just wants to have some *fun*, and she doesn't care who knows it.

A SONG TITLE	AN ALBUM	A BAND	A LYRIC
"Freedom at 21" – Jack White (2012)	*Rick Is 21* – Rick Nelson (1961)	Trisome 21 – Lille, France, goth/dance (1980s-1990s)	"Twenty-one drums and an ol' bass horn/ Somebody beatin' on a ding-dong" – Roy Hall, "Whole Lotta Shakin' Goin' On" (1955)

LAST WORDS:
"Twenty-one! The Chief's vote makes it twenty-one! And by God, if that ain't a majority, I'll eat my hat!"
—Randle P. McMurphy, in Ken Kesey's *One Flew Over the Cuckoo's Nest*

David Klein

"I live in an apartment on the 99th floor of my block"
– Rolling Stones, "Get Off of My Cloud"

"Elevator! Going up!/ In the gleaming corridor of the 51st floor/ The money can be made if you really want some more" – The Clash, "Koka Kola"

"When they're droppin a piano from the 47th floor/ I'm the guy underneath 'em looking up"
– Lovin Spoonful, "Pow"

"Number 36!" – Van Morrison, "Cleaning Windows"

"On the thirty-first floor a gold-plated door
Won't keep out the Lord's burning rain"
– Flying Burrito Brothers, "Sin City"

"Let's explore/Up there, up there, up there on the twenty-fifth floor" – Patti Smith Group, "25th Floor"

"She lives on the 20th floor uptown/the elevator's broken down…" – Eddie Cochran, "Twenty Flight Rock"

"There's a light on the nineteenth floor tonight"
– Game Theory, "Throwing the Election"

"18th Floor Balcony" – Blue October

"On the Seventeenth Floor" – Living Death

"Song From the Sixteenth Floor" – Paul Kelly

"Don't get my sympathy hanging out the 15th floor" – Radiohead, "Just"

"And catch the grey men when they fall from the fourteenth floor" – Steely Dan, "Black Friday"

"13 Stories High" – Tonto & the Renegades

"12th Floor Vendetta" – The Damned

"I'm up on the eleventh floor watching the cruisers below" – "Queen Bitch," David Bowie

"10 Storey Love Song" – The Stone Roses

"Cats can fly from nine stories high.
And pigs can see"
– Johnny Cash/The Highwaymen, "Death and Hell"

"A bird sits on a branch of my eighth-floor window"
– Regina Spektor, "8th Floor"

"Heaven on the Seventh Floor" – Paul Nicolas

"From a sixth-floor window a gunner shot him down"
– The Byrds, "He Was a Friend of Mine"

"Can't stop the fire (can't stop the fire) /
we're living on the fifth floor and it's okay"
– St. Etienne, "Fifth Floor"

"Dropped four flights and cracked my spine"
– Grateful Dead, "Tennessee Jed"

"Third Floor Heaven" – Be Bop Deluxe

"My name is Luka/ I live on the second floor"
– Suzanne Vega, "Luka"

"Ground floor/ ladies clothes, sportswear, stationery"
– Ladytron, "Paco!"

22

If 21 is all about the thrill of being deemed a full-fledged adult in the eyes of the law, 22 kills that exuberance dead. "22" is often sung in a noticeably dark spirit, like Bo Diddley's "Tombstone hand and a graveyard mind/I'm just twenty-two and I don't mind dyin'"[1] or "Lost Highway," a song by blind musician Leon Payne made famous by Hank Williams, whose antihero first strays when "just a lad, nearly 22/Neither good nor bad, just a kid like you." David Bowie's late-period "Love Is Lost" echoes the theme: "This is the darkest hour/you're 22/the voice of youth/the hour of dread." It's curious that such a young age is so freighted with unsettling implications. Why does Paul Simon sound so dire in "Leaves That Are Green" when he sings, "I was 21 years when I wrote this song/I'm 22 now but I won't be for long"? Billy Bragg's "A New England" begins with the same couplet. And the concept is not restricted to introspective folk-rock either; in "Gangsta's Paradise," Coolio raps, "I'm 23 now, but will I live to see twenty-fo'?"

> "That's some catch, that Catch-22," he observed.
> "It's the best there is," Doc Daneeka agreed."
> —Joseph Heller, *Catch-22*

It would be a tall order to capture the absurd, circular logic of Joseph Heller's *Catch-22* in a song.[2] But the term itself has proven useful to songwriters of all stripes. Pink, Billy Squier, Peabo Bryson, Chris Spedding, Yngwie Malmstein, and second-wave English punkers Infa Riot all have songs called "Catch 22." The Specials name-checked "catch-22" memorably in their 1979 debut single, "Gangsters." Concurrently, Paul Weller and his Jam began their domination of the English singles charts with "Eton Rifles." The Jam is long gone, but Weller continues to be a force. The title track from his 2008 collection, *22 Dreams*, finds him spinning out a tight soul-revue rave-up animated by his own wailing guitar and the kinetic groove of drummer Steve White.

Lily Allen (who was born in 1985, the same year Weller sang on the "Do They Know It's Christmas" mega-single) wrote "22" in 2007, when she was 22. Like Weller's work, Ms. Allen's songs are as British as fish and chips. Here, her vocals hark back to a prewar music-hall style as she wonders where a self-styled party girl like herself might end up by the time she reaches the lofty age of 30, when "society says her life is already over." No comparable doubts creep into Taylor Swift's "22," a profoundly infectious 2012 single that commemorated what she called "my favorite year of my life."

---

[1] In Richard Thompson's tour de force "'52 Vincent Black Lightning," Thompson, perhaps unconsciously, echoes Diddley's line: "And now I'm 21 years/ I might make 22/ But I don't mind dyin'/ but for the love of you." Erika M. Anderson, operating as EMA, consciously echoes the Diddley lyric in the vocal ad-libs at the end of "California" (she also interpolates "Camptown Races"). See also: "Hold On" by the Alabama Shakes: "Didn't think I'd make it to 22 years old"

[2] The Grateful Dead loved Vonnegut enough to name their music publishing company Ice Nine, after the pernicious chemical compound from *Cat's Cradle*.

David Klein

Reaching the age of 22 is at best a bit of an anticlimax following the legal emancipation of 21, but many songwriters have employed the numeral to signify a certain precipice between the teen years and actual adulthood. As we learned from Baz Luhrmann's "Everybody's Free (To Wear Sunscreen)," based on lyrics famously misattributed to Kurt Vonnegut, "the most interesting people I know didn't know at 22 what they wanted to do with their lives." Neil Young enlivens that sentiment far more succinctly in "Powderfinger" (And I just turned 22/I was wondering what to do").

The Flaming Lips' aggressively psychedelic "When Yer Twenty Two" has a perpetually falling-forward rhythm that seems to mimic existential awkwardness. Wayne Coyne's urgent vocals anchor it, spinning out oblique lines that verge on the sensical. At 3:10 there's a near-subliminal sonic zoom, from left speaker to right speaker, which has no doubt blown some minds over the years. "When Yer Twenty Two" captures the roiling peak of the drug experience, where the senses merge, consciousness takes on a new shape, and lines like "Eggs break when you walk on the scramble" are perfectly comprehensible.

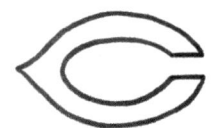

22 WORDS ABOUT "TWENTY-TWO IN CINCINNATI" BY MARTHA & THE MUFFINS:
Early production work by the young Daniel Lanois in this cavernous, dub-style instrumental from 1981. Originally a B-side. Miles from "Echo Beach."

Guaranteed to send all but the most intrepid of acid trippers fleeing to a more secure location is "22 Going on 23" by Butthole Surfers, the final track on 1987's *Locust Abortion Technician*.[3] This slab of sample-happy mayhem begins with the voice of an addled radio-show caller: "Last July I was assaulted, sexually, and ever since then I've had trouble sleeping." The looped vocal snippets ("I can't sleep, I can't sleep"), syrup-thick bass line, and final bovine chorus may have been conjoined in a spirit of wacky fun, but together they add up to a deeply unsettling track.

The eternally unsettling *Blue Velvet* made memorable use of "In Dreams," the classic Roy Orbison ballad, and Orbison's "Twenty-Two Days" has some of that song's earmarks. The emphatic strings, the cushion of cooing chorus, and of course, Roy's magical voice are all there, but the song, written by the oft-times treacly Gene Pitney, feels less than top-drawer for an established master.

The same could be said, and in fact it has, about the later work of Nina Simone. Those who first encountered Ms. Simone via "My Baby Just Cares For Me" in a TV ad for Chanel No. 5 will find themselves utterly unprepared for "22nd Century," a 12-minute prophesy sung-chanted over an unchanging piano vamp, in which Simone declaims lines like, "Prevention of employ to animals/Flying things, revolutions of music." In a Salon review of a Simone compilation that leaned heavily on her weirder work, Tony Scherman wrote simply, "The last song, '22nd Century,' is the worst thing I've ever heard in my life." For that alone it's worth checking out. How can you not be curious?

In Nina's prophesy, the early '90s riot grrrl scene would not have happened ("1990 was the year when the plagues flood the earth"). But somehow, one imagines, she would have appreciated, possibly even covered, Babes in Toyland's "22," a brutal pounder that incorporates exotic minor-key motifs à la Dick Dale, along with beguiling glossolalia ("Fachea anima interstellar/Hidden in logorrhea"). Kat Bjelland utters the title numeral at the very last second, in a fierce whisper.

Bruce Dickinson, on the other hand, spits out the title address of Iron Maiden's "22 Acacia Avenue" with typical heavy metal venom. The song's subject, Charlotte the harlot, appears in no less than four Iron Maiden songs. According to Maiden main man Steve Harris, Charlotte was based on a real person. Was she a prostitute? "Actually," Harris said in an interview, "she was more of a

---

3   A handwritten list of Kurt Cobain's 50 favorite LPs, unearthed in 2012, included *Locust Abortion Technician*, the Shaggs' *Philosophy of the World*, Bad Brains' *Rock for Light*, and *Meet the Beatles*.

slut … I mean, if you turned up at her house with some booze or some speed, you were more or less guaranteed a lay."

Ike Reilly's "22 Hours of Darkness" is an impassioned litany of things that are "not enough"—from clothes and teachers to friends and family. And it keeps returning to a blunt question: "Do you think that maybe this is just your life?" Reilly starts off quizzical, but he builds it up, brick by brick, to a pulse-pounding finish, reeling off his layered rhymes in a rusty tenor that has real bite. "22 Hours" lays out one of rock 'n' roll's basic truths: You don't have to stand for the way things are.

THE VERDICT

Roughly four years into the new millennium, the British indie rock world was briefly and somewhat inexplicably glutted with bands all hyped up on Wire, XTC, Gang of 4, and their tense, angular post-punk brethren. Franz Ferdinand were on the leading edge, the most likely to have staying power, while the burly middle swath—Maximo Park, Kaiser Chiefs, Dogs Die in Hot Cars, Bloc Party, all fronted by a variety of yelpers—had similar influences and could easily be mistaken for one another. And they all had some arresting music in them.

The Rakes, a smartly dressed London-based quartet, launched themselves to chart-toppery with the sizzling "22 Grand Job." Their sound may have been less overtly indebted to the past than the aforementioned acts, but they were certainly of the scene, touring with Franz Ferdinand and hobnobbing in the *NME* tent on the festival circuit. "22 Grand Job" combines a blunt live-band sound with the quick-shifting precision of a gymnast working the uneven parallel bars, Alan Donahoe's half-spoken "it's awroight, it's awroight" making for a slyly insinuating refrain.

An instant hit, the song generated genuine excitement. Thus softened, the charts were conducive to several more excellent Rakes singles, although one could argue that none had the focused impact of "22 Grand Job." The Rakes didn't last very long, and neither did the moment. After a few years, supply caught up with and exceeded global demand for the second coming of a sound that was never very commercial to begin with. After relocating to Berlin for a fresh start on their third LP, the Rakes, not wanting to beat a dead horse, abruptly called it quits on the eve of a 2009 tour. The job was done.

A SONG TITLE	AN ALBUM	A BAND	A LYRIC
"22" – Night Beds (2013)	*Log 22* – Bettie Serveert (2003)	22 Band Kankan – Guinea, pop (1980s)	"22 faces, disintegrating" – Siouxsie and the Banshees, "Christine" (1980)

23

Believers in the 23 Enigma contend that this number connects everything. William Burroughs believed, and he found support both in the empirical (the 23 chromosomes that join together to create life itself) and in legend (23 is the number of the Illuminati, the secret cabal believed by conspiracy theorists to control world events). According to the author-philosopher Robert Anton Wilson, Burroughs also considered 23 a flat-out "death number." Wilson writes that Burroughs was apt to cite Psalm 23[1] (a staple at funerals), the 23 knife wounds dealt Julius Caesar, and Bonnie and Clyde's death day, May 23, 1934, as juicy examples.[2] So, is 23 a magic number?

Blonde Redhead addresses the 23 Enigma head-on ("23 seconds, all things we love will die/23 magic, if you can change your life") on the title track from *23*, a record that found the band toning down its noisier impulses in favor of traditional song structures and, as here, the occasional la-la-la chorus. That it's all set against slaloming *Loveless* guitars transforms these elements into something rich and strange, with Kazu Makino's vocals recalling the celestial intonations of Miki Berenyi of '90s shoegaze pinup band Lush. Appropriately enough, the band's later records have been issued on Lush's old label, 4AD. Suffice to say, "23" would bring a wry smile to the gaunt visage of William Burroughs.[3]

> **Max Bialystock:** "We are seeking Franz Liebkind."
> **Concierge:** "Oh … the Kraut! He's on the top floor, Apartment 23."
> —*The Producers* (1968)

There is no agreement on the origin of the phrase "23 skidoo." In some late-19th-century staged versions of *A Tale of Two Cities*, Sydney Carton was the 23rd person to be beheaded; some have contended that this instance led to the term 23 skidoo in a parody version. But other explanations abound, ranging from a children's jump rope rhyme to the 23rd Street address of the Flatiron Building in Manhattan, where police would use the operative phrase to direct the hordes of men gathered there to disperse. According to Merriam-Webster, the skidoo part simply derives from "skedaddle," meaning a quick departure.

Whatever its origin, the phrase inspired the naming of a critically lauded English outfit from the early '80s. 23 Skidoo emerged from the post-punk industrial scene with a tone that was less confrontational than that of noise-ateers such as Throbbing Gristle[4] and the Pop Group. The band's few records did not make much of an impact except on crate-digging connoisseurs like

---

[1] The actual Dead (as in Grateful) invoke it in "Alabama Getaway" from 1980's *Go to Heaven*: "Twenty-third Psalm Majordomo/Reserve me a table for three/In the Valley of the Shadow/Just you, Alabama and me"
[2] There are exceptions, of course, such as the 23 passengers who survived the Hindenburg's final transatlantic voyage.
[3] Burroughs's dystopian visions influenced industrial rockers like Ministry, Skinny Puppy, and the Belgian band Front 242, a member of which goes by Richard 23.
[4] When Genesis P'orridge issued *LIVE IN TOKYO* with his post-Gristle outfit, Psychic TV, the liner notes stated an intention that would have pleased William Burroughs: "There will be 23 live albums. One released every month for 23 months on the 23rd of each month." In no time, though, the schedule proved too daunting to live up to.

the Chemical Brothers, who were so enamored of 23 Skidoo's jaggedly funky "Coup" that it pretty much serves as the basis of the bros' hit "Block Rockin' Beats."[5] Sixteen years after the release of 1984's *Gamelin*, 23 Skidoo put out a new record, which included "Catch 23," a title that in retrospect seems inevitable.

Luna's languid "23 Minutes in Brussels" stretches out like a lazy day. With a title inspired by a similarly named recording of a notorious concert by proto-punks Suicide, the song opens with a seeming "Proud Mary" allusion ("Left my hotel in the city"), Dean Wareham's earnest tenor at odds with the uneasy portent of the refrain ("Say a prayer for you and me"). The song proceeds at a pace both laid-back and determined, augmented by a gorgeous extended middle section featuring the signature six-string squalls and squiggles of Tom Verlaine[6] set against Wareham's liquid silver tone. In his memoirs of the indie rock life, *Black Postcards*, Wareham reports that "23 Minutes" was the last song Luna played at its final gig. He also confirms that he has no love for Brussels. On a not-unrelated note of foreboding is "23 Lies" by Death in Vegas, a slow throb permeated by the narcotic soprano exhalations of vocalist Susan Dillane. These two would go well on a mixtape created by the dead twins in *The Shining*.

THE VERDICT

Sometimes a song so owns a number that it can even fend off an exhilarating newcomer like Blonde Redhead's "23" or a stalwart with the effortless cool of Luna's "23 Minutes in Brussels." "Strawberry Letter #23," written by the guitar prodigy Shuggie Otis and turned into a hit by the Brothers Johnson in 1976, is such a song.

Oh, it's certainly less 23-centric than the others—the numeral 23 is never even uttered, although 22 is, twice. And admittedly the Johnsons did not exactly reinvent the original. They stick pretty close to Shuggie's version, from that central insinuating hook answered by ascending "ooh-hoo-hoos" to the spacey Steve Miller-ish interlude. Yet the Johnsons' version is definitive beyond any doubt. The vocal melody has been hammered into shape, and the backing vocals add exquisite color. But more than the skilled producer's hand of Quincy Jones, the main thing that Otis's version lacks is the supple, popping bass of Louis "Thunder Thumbs" Johnson, which transforms an ethereal serenade into dance-floor dynamite. The bros' panache in the packaging department was similarly top-tier; the 12-inch single of "Strawberry Letter #23" was released on red vinyl with a strawberry-scented sleeve.

A SONG TITLE	AN ALBUM	A BAND	A LYRIC
"23 Beats Off" – Fugazi (1993)	*Dear 23* – The Posies (1990)	Demolition 23 – New York City, punk (1993–1995)	"I wanna grovel in mud/ And sing the 23rd Psalm" – Chris Knox, "Wanna!!" (1989)

---

5 Liquid Liquid, compatriots of 23 Skidoo on the other side of the Atlantic, had a similar thing happen with their song "Cavern," which was pretty much grabbed wholesale for "White Lines" by Grandmaster Flash and the Furious 5.
6 Verlaine's "Words From the Front," from the 1982 LP of the same name, begins, "January 23rd … there's no road…it's been raining now for three days, we're in mud up to our knees …"

Certain numbers have very specific associations, none more so than 24. References to 24 in song almost always deal with the critical division of time equivalent to a full day. There are exceptions, of course, like Lambchop's "Jan. 24," which reimagines Ray Charles's "What'd I Say" as a new wave surf instrumental. Surely no one has ever enunciated the numeral with more conviction than Muddy Waters on "Twenty Four Hours" when he growls, "She been gone twenty-four hours/and that's twenty-three hours too long."

Mudhoney, in the sludgy, hyperactive "Twenty Four," have thoroughly micromanaged their next 24 hours, and the schedule is (six) packed. On "24 Hour Party People," Happy Mondays advocate a similar level of recklessness, albeit with the unspoken addition of powders and pills. The song gave an excellent film its name, and Sean Ryder's ragged vocal cords sound like they've already been through the grinder in this would-be anthem to transcendent excess.

Joy Division, on the other hand, whose beginnings were also depicted in that film, gives us "Twenty Four Hours." For Ian Curtis, suffering from depression, personal stress, and seizures, a day was something to endure, and the song captures the tight grip of raw dread. There's no bridge; the music ebbs and flows like an angry ocean. Curtis's unearthly baritone floats above it, issuing some of the most nakedly honest lines in a brief career marked by naked honesty: "Now that I've realized how it's all gone wrong/Gotta find some therapy, this treatment takes too long."

Gene Pitney isn't remembered as well as Roy Orbison and Del Shannon, but he belongs squarely in that same tradition of the emotional balladeer whose songs limn the landscape of heartache. In "Twenty Four Hours From Tulsa," the singer meets his dream lover at a small hotel that's just a day's drive from where his baby awaits. Complications ensue. "Tulsa" is vintage Burt Bacharach-Hal David, with mariachi trumpets, hooks galore, and a tasty story to tell. Pitney also recorded "24 Sycamore," a heart-on-the-sleeve wailer that is the dark side of "Twenty-Four Hours From Tulsa." Now the singer is in a hell of his own making. He ditched the wrong girl, and he knows he's doomed to forever pine for the hand he used to hold at ... you know where.

THE VERDICT

The critics used terms like power pop, jangle pop, and the Paisley Underground, but the labels weren't particularly useful to me; I just knew that by 1986 or so, there was a bunch of bands that tended to lump loosely together in my head (with R.E.M. at the center), based chiefly on Big Star/Byrds-like guitars and melodies and/or earnest, Alex Chilton-esque vocals. The Three O'Clock, the Long Ryders, the Windbreakers, Tommy Keene, Let's Active, 28th Day, the Bangles: I loved them all. My absolute faves were

Game Theory, who added a level of literary sophistication to the power pop enterprise. With song titles composed of Kubrick movie quotes, album covers sporting enormous fonts and cryptic letter-number sequences, and lyrics that made reference to highbrow books and lowbrow culture, Miller's sometimes-manic displays of ingenuity impressed me, as did the guitars, even if I couldn't always fully comprehend what was being said.

Game Theory's "24," the lead track from 1985's *Real Nighttime*, is among the band's more instantly fathomable works. Unlike any other song under consideration, "24" is about *being* 24,[1] and it nails, in oblique terms, the age's particular "where am I going, where have I been?" ethos ("And everything is in terms of next time/Twenty-five thousand more miles to the dateline/Is it because I'm 24, not 25?"). Twenty-four is the last time you'll be closer to 20 than 30. It's the last good look you'll have of your teenage years, because pretty soon a fresh numerical truth will come into view as childhood recedes into the distance. At more than twice 24, I still find much to admire about "24": the sweet, folky sound of the acoustic guitars, the rainbow-bright sheen of Mitch Easter's production, the yearning melody, the economy (2:49), and the proud declaration of the title number. Oh yes, and those wispy vocals.

A SONG TITLE	AN ALBUM	A BAND	A LYRIC
"Twenty Fourth Hour" – The Action (1967)	*24 Carat Purple* – Deep Purple (1975)	24-7 Spyz – Bronx, N.Y., funk/hard rock (late 1980s)	"Twenty-four before my love, you'll see/ I'll be there with you" – Yes, "Roundabout" (1972)

AFTERWORD

A lot has happened since I originally wrote these words in 2010 or so. Scott Miller died, suddenly, in spring 2013, at 53. Hearing about his passing gave me the peculiar twinge of regret I imagine people experience when they learn of the death of a loved one with whom their last encounter was an argument. It wasn't that we'd argued, or even that we knew each other personally. The feeling stemmed from my attempts to reach him in regard to this book. I'd written to him a few times, hoping for some response, some commentary, some something. Maybe I just wanted him to know that the book existed, that I existed, and that I loved his music.

I'd fallen hard when I first heard Game Theory. Hell, sometimes I'd go through the longboxes in the G section at Tower Records just to see if there was a record of his I'd somehow missed. Scott Miller sang in a precise but gentle tenor that signaled vulnerability. Nevertheless, in the production and in Miller's words—the postgraduate lyrics and even the use of text on the band's LP jackets—Miller's own baroque sensibility was always on display. Album titles riffed on John Cheever (*The Big Shot Chronicles*); a famed double-play combination (*Tinker to Evers to Chance*); a line from a cheesy America song (*Plants and Birds*

---

[1] Neil Young weighs in on "Old Man" ("Twenty-four and there's so much more"), and, while not technically a #24 song, Stephen Stills' "4 + 20" (from Crosby, Stills, Nash, Etc., Etc.'s *Déjà Vu*) is a chilling narrative told by a 24-year-old who finds himself "just wishing that my life would simply cease." When I was a kid, a teenage babysitter named Andy Feffer played this song for my brother and me—but first he insisted that we sit down and hush up so we could really take it all in. Thanks, Mr. Feffer, wherever you are. You made a powerful impression on me by insisting that sometimes you really have to *listen*.

and Rocks and Things). And the girls, those mysterious evanescent girls he sang about: "We Love You Carol and Alison," "The Real Sheila," "Like a Girl Jesus."

Somehow it was never just a massively clever guy at work. Miller was high-minded but never high-handed. Anyone who credits himself with "Guitar, miserable whine" does not suffer from excessive self-regard. A romantic heart surged through these wonderfully melodic, carefully wrought songs.

Still, even as I considered the music and all it had meant to me, there was that nagging association with my petulance over never hearing back from him. Of course death has a way of erasing hurts both big and small. Petulance erased, I remembered something. In late 2000, I had posted a message on the website of Miller's post-Game Theory vehicle, the Loud Family, in a section called Ask Scott. I'd vented my frustration about not being able to find *Attractive Nuisance*, the band's latest release, in any New York City record store. When Miller died, I had only a dim recollection of what he'd written back, so in a strange way, reading his note to me was like reading it for the first time. It was as if he'd responded after all.

"*Thank you for writing. So many things didn't quite click in my music career that no particular one irritates me anymore. What I do is somewhat inherently uncommercial (both my content and my not overly obvious vocal merit), and when I look back I'm a little astounded that so many people supported me. It's weird to reflect that there was a time when I would walk into a record store in London and actually be recognized — a memory that seems oddly parallel to going into the same store as a teenager and being in awe of anyone who had a record on sale there.*

So I'm thinking of everyone who bothers to read this on Thanksgiving.
come on pilgrim,
—S

25

Twenty-five is often associated with anniversaries, whether matrimonial or historical, but no one ever wrote a great song about a 25th anniversary. What matters to songwriters is that 25 is a melodious sounding numeral that plays well with others, in the Mott the Hoople "Speed jive/don't want to stay alive/when you're twenty-five" sense of things.

Disqualifiable on technical as well as aesthetic grounds, "In the Year 2525 (Exordium and Terminus)," by Zager & Evans, is excruciating on many levels. Here's just one: when Z&E have completely exhausted the thesaurus for words that rhyme with five, they blithely begin singing about the year "sixty-five ten." Shameless. A far more palatable oldies-radio staple is Chicago's urgent "25 or 6 to 4." Based on a memorable descending hook, the song has an admirably numerical title, albeit one that has puzzled listeners for decades. It was thought to be some kind of mystical drug reference, but "25 or 6 to 4" apparently refers to the early-morning hour when the actual time can only be approximated. With its charged-up, undeniably beefy central riff, this is perhaps the only Chicago song that would lend itself to a satisfying emo cover.

Chicago (the city, not the band) was home to Veruca Salt (the band, not the *Charlie and the Chocolate Factory* character). Their debut, *American Thighs*, included three number songs, and on "25" Nina Gordon demonstrates a solid understanding of basic math facts. Johnny Cash was also fond of counting songs. In his version of Shel Silverstein's "25 Minutes to Go," he counts down the moments leading up to his execution over a slowly building hoedown beat for the inmates at Folsom Prison (not to mention "One Piece at a Time," which ends with Cash's peppy recitation of numbers 49 through 70).

An obvious contender for #25 glory is "Twenty Five Miles" by Edwin Starr. Combining a counting-down lyrical structure with an irresistible Motown jam, this was a major crossover hit in 1969. While not quite as killer as Starr's "War" (in which he delivers what has been called—by me—the definitive "*huuuh*" in recorded music), this journey song is spurred on by a driving beat and the stirring Starr vocals. I've always been ambivalent about the fact that he never gets home though. With five miles to go, there is a final self-exhortation of "I got to keep on—*walkin*" and some seriously funky drumming before the song fades. I guess it would have been lame to end with "I got 60 more feet to go now, I got 50 more feet to go…," but the song still feels somehow unresolved. Most listeners will assume that Starr makes it home, and I guess I can live with that.

"25th Floor" comes from the Patti Smith Group's 1978 commercial breakthrough, *Easter*, which earned Smith a huge hit with the Springsteen-penned "Because the Night." Despite the move to the mainstream, abetted by the canny production instincts of Jimmy Iovine, *Easter*'s bedrock of primal riffs and polymorphous poetry is a slam-dunk. Killer rock couplets and free verse exhort us to break free of our bonds, to shine like lotus blossoms rising from the muck. And it *rocks*. The end-of-song soliloquy might not approach the majesty of "Why must we pray screaming/why must not death be redefined?" from "Dancing Barefoot," but that would be asking a lot.

Two drug songs released in the same year vie for top honors here. "25 O'Clock," by XTC's alias, the Dukes of Stratosphear, opens with a slowly gathering cacophony of clock sounds. From there it makes musical reference to a half-dozen or so proto-acid rock songs, from the Amboy Dukes' "Journey to the Center of the Mind" to the Electric Prunes' "I Had Too Much to Dream Last Night," all filtered through XTC's blender consciousness. *25 O'Clock* was something of a tongue-in-cheek affair; the band members assumed psychedelic noms de rock, like Cornelius Plum and the Red Curtain, and did nothing to link themselves to the EP when it came out on April Fools' Day 1985. But the music is too heartfelt and ingeniously well played to be written off as mere pastiche. Produced by the legendary John Leckie (The Stone Roses, Radiohead, George Harrison, Pink Floyd—legendary enough for you?), these five songs can be heard playing on the jukebox in an alternate universe where flower power never wilted.

That jukebox would also include "25 Pills" by 28th Day, a numerically named trio from Northern California who were lumped in with the Paisley bands of the mid-to-late '80s. Although venerable indie rocker Barbara Manning is clearly the star of the band, on "25 Pills" she steps back and harmonizes with Cole Marquis's lead vocal, for a bracing blend of male/female voices that recalls the spirit of such vintage pairings as John Doe and Exene Cervenka or even Grace Slick and Paul Kantner of Jefferson Airplane. For the final chorus, Marquis reaches deeper and flies higher, and it feels like napalm in the morning, like victory. "25 Pills" is "I'm Waiting For the Man" Northern California-style. Sure, they want to get high; they're just in a little less of a hurry about it.

THE VERDICT

Doing an end run around this trippy transatlantic battle and earning top honors is "25 Lighters," an influential hip-hop song by Port Arthur, Texas, deejay/producer DJ DMD and rappers Fat Pat and Lil' Keke. Built upon a rubbery beat constructed from an Al B Sure sample, the song leverages the easy-flowing syllables of twenty-five for all they're worth. Not only is the number essential to the hook, it's the linchpin of the verses as well, setting up a street-centric litany that just makes you want to shake your ass. The indelible hook comes when Lil' Keke answers the choral refrain of "25 lighters on my dresser, yes sir" with the beguiling and perfectly completing response, "*I gots to get paid.*"

As opposed to the other drug-oriented songs in contention, this one takes a distinctly non-recreational perspective. Those lighters on the dresser have been disassembled and stuffed with crack cocaine, ready to be hawked on street corners in plain sight. At least that's the prevalent interpretation. Yet that wasn't necessarily the intent of its creators. The phrase "25 lighters" had cropped up on a 1995 song called "All in My Mind" by 8Ball and MJG. In that song, the lighters merely served as ignition devices for the "two hundred eighty pounds of hay" in MJG's possession, hay meaning weed.

The phrase first caught the ear of Dorie Dorsey, aka DJ DMD, in 1998 when he reassessed some verses by rapper Lil' Keke from a 1996 session. Dorsey was so taken with the line that he made it the central hook of a new beat, to which he added one verse by Fat Pat, who was murdered in early 1998 (it was never solved), and added one of his own. In doing so, he took those lighters from a secure apartment stacked with herb and into a world of playas, haters, and gats. It's hard to interpret a line like "on the 'vard is where I sling" and numerous references to "the game" as anything but references to the drug trade. Indeed, the official version on Answers.com advocates for the crack-filled lighters interpretation. It's also the version Billy Gibbons of ZZ Top tells about the song that haunted him until he finally covered it in 2012. Yet that same year, speaking to *Texas Monthly* reporter Michael Hall, Dorsey took no credit for the intent behind his touchstone image. "The phrase 'twenty-five lighters' didn't mean anything to me ... I really didn't do drugs," he said. "I just liked the sound of the number twenty-five. [Aha!]

In the ensuing years, DJ DMD has been through ups and downs and found God. Although his song is now a well-loved classic, sampled by the likes of Z-Ro, Big K.R.I.T., and Kendrick Lamar, it wasn't exactly paying bills or bringing him wide acclaim. That is, not until he got a call from Billy Gibbons's people, seeking permission for the guitarist to remake the song as a swampy groover called "I Gotsta Get Paid." Gibbons offered to share songwriting royalties with him. Dorsey was flabbergasted. Then the royalty checks started showing up.

Dorsey still loves beat making. Moved to rerecord his signature song, he found a way to reflect his current clean-living perspective in a track he renamed "#25BiblesOnMyDresser." That old hook line, which always sounded like a spirit voice, now responds: *"I gots to get saved."*

A SONG TITLE	AN ALBUM	A BAND	A LYRIC
"Twenty Five Years" – Saint Etienne (2012)	*25 to Life* – The Reverend Horton Heat (2012)	Section 25 –England, post-punk, (1980s)	"We got twenty-five rifles just to keep the population down" – Neil Young, "Revolution Blues" (1974)

26

But for marathon runners and aficionados of rhombicuboctahedrons (26-sided solids), 26 doesn't crop up too often. There are 26 letters in the English alphabet (Philip Roth dubbed it "the Big Twenty-Six" in *The Great American Novel*), and there are 26 red cards and 26 black cards in a standard deck, but the number doesn't hold a place in our hearts. Shel Silverstein, who gave the world "A Boy Named Sue" as well as *The Giving Tree*, seems to have tapped the numeral for its very flimsiness in his "26 Second Song," a goof country ballad that makes a quick dash for home to meet its self-imposed time limit. After surveying the field, I've concluded that 26 is not a quantity that often stirs the soul of a songwriter—the 26 perfectly modulated "I knows" loosed by Bill Withers in "Ain't No Sunshine" notwithstanding.

Lou Reed might well have been thinking of the meter in "Waiting For the Man," when he sang, "26 dollars in my hand." But it's curious that he chose 26, rather than say, 25, which has the same cadence and makes more sense.[1] Seriously, what kind of dealer charges $26 for a bag? On a related note, Cathy from Jim Carroll's "People Who Died" was 11 when she pulled the plug on 26 reds and a bottle of wine.

Maybe 26 deserves better. I thought Chic could pull it off, and Messrs. Edwards and Rogers certainly tried. In "26," they attempt to coin a novel usage, but the result is surprisingly clumsy in the execution. Sorry, but "My baby's a 26 on a scale of 1 to 10" lacks any semblance of the streamlined disco-tastic-ness of Chic at their best.[2] Chic's full vindication would take a few decades' worth of rampant sampling and ancestor worship to occur. From this vantage, it's clear that Chic is to dance music what Chuck Berry is to rock.

Certain brave souls have deemed 26 a good, evocative numeral to use in song. The Pretty Things engage in a bit of sloganeering on "October 26" (seemingly inspired by Russia's October Revolution, probably also the case for Beulah's "Comrade's Twenty-Sixth"). This languorous concoction from *Parachute* finds the scruffy-sounding Pretty Things trading in earthly concerns for hippie platitudes ("Revolution is all that it can be"). But lovely harmonies and a true mature-Beatles sound via producer Norman "Hurricane" Smith make it more palatable than your average pabulum.

A definite pattern begins to develop at 26 in the birthday realm, where reaching a #_6 birthday (e.g., 36, 76) nudges you inexorably closer to an age that once seemed unreachable. At 16, in anticipation of the age of 20, this is a good thing, but beyond that it fills any sensible person with existential dread. Turning 26 is an ungainly, even ungodly thing, depending on the birthday boy/girl's state of mind. In a quotation misattributed to Margaret Thatcher, presumably because it just seems so *apt*, the Iron Lady is purported to have said, "A man who, beyond the age of 26, finds himself on a bus can count himself as a failure."

It's not clear what Maggie would have said about a woman, such as the title character in Pavement's "Robyn Turns 26," who finds

---

[1] Reed had a fondness for crooked numbers, notably, "You're number 37 have a look," in "Femme Fatale,"; "You couldn't get to page 17" in "Kill Your Sons," and "You spit on those under twenty-one" in "European Son." Interestingly, all three usages are accusations.
[2] Jackson Browne handily achieved poetry with his take on this type of hyperbolic locution in "Redneck Friend," a masterstroke of double-entendre: "Cause he's the missing link/the kitchen sink/eleven on a scale of ten."

herself in the same position. The song nevertheless echoes a certain Thatcherian lack of empathy for the struggles of common folk. Sampling an Indian rain dance chant for comedic effect, interpolating a Night Ranger hit, and spotlighting the limited rap abilities of Stephen Malkmus's then-girlfriend, "Robyn" has piss-take written all over it. But even with a one-off, Malkmus is a skillful skewerer of the very people inevitably attracted to his music, and his line of speculation ("Got 20 Camel Lights, a six-pack of brew/That's 26 friends") is both hilarious and cruel at the expense of the vapid Robyn.

In Stereolab's "Olv 26," Laetitia Sadier trills Marxist slogans in French and English as a synchronized pulse slices through a rumbling synthetic squelch. But what might in less humane hands be robotic and dull, Stereolab makes alluring and playful. Changes of texture and timbre over a mostly unchanging structure—that's what you want in a Stereolab song, and what they deliver here. As far as I can tell, Olv signifies "our lady victory," but I am far too dense to comprehend lyrics like, "Depuis le temps que c'est promis nous irons tous au paradis," which appear to be in some kind of foreign language.

THE VERDICT

The Four Preps were a well-scrubbed West Coast vocal group remembered as much for being inspirational to Brian Wilson of the Beach Boys as for their dreamy harmonies. Their biggest hit was "26 Miles (Santa Catalina)," which rose to No. 2 in 1958, edged out of total domination by the Champs' "Tequila."

The song had a long gestation period. As a 15-year-old, the group's lead singer, Bruce Belland, composed the chord sequence on a ukulele he was given to pass the time while recuperating from a bike accident. The 26 miles part came some time later, while bodysurfing with a friend who musingly gauged the mileage to the distant but visible island of Santa Catalina. As it turns out, his friend was a bit off the mark. Belland told the *Los Angeles Times*, "It's really like 22.3 miles, but you try singing that. Think about that meter!"

Belland enlisted the help of Glen Larson,[3] the group's baritone, to finish the song, but their handlers at Capitol Records didn't rate it too highly and tapped it for a B-side. When the vocals were too low in the initial recording, the Preps were told to simply sing over the original mix. The resulting aural effect is part of what makes this wind-tickled tropical fantasy so evocative. "It gave the song this washy sound," Belland said. "It was like you were hearing the song echo in an underwater grotto."

A SONG TITLE	AN ALBUM	A BAND	A LYRIC
"Pattern 26" – Manifesto (1992)	*26 Mixes For Cash* – Aphex Twin (2003)	Apartment 26 –Leamington Spa, England, alt-metal (1990s–2000s)	"26 vanilla swarm" – Kate Bush, "50 Words For Snow" (2013)

---

[3] Larson, who died in late 2014, went on to an extremely successful career in television, creating *Battlestar Galactica*, *Magnum P.I.*, and *Knight Rider*, as well writing the latter's oft-sampled theme music. Belland, a successful songwriter after the Preps disbanded in 1969, also made his way to TV, as a writer and voice artist. He still performs with a retooled Four Preps lineup. Original member Ed Cobb, who bailed in '66, penned the classics "Tainted Love" and "Every Little Bit Hurts."

27

> "It was about then [1920] that I wrote a line which certain people will not let me forget: 'She was a faded but still lovely woman of twenty-seven.'" —F. Scott Fitzgerald, "Early Success"

By the age of 27 you are supposed to have learned and experienced enough to become, more or less, the person you are going to be for the rest of your life. When you mess up at 27, you can't blame it on being young and stupid anymore. Billy Beane, the ex-MLB wunderkind player whose transformation into wunderkind GM is documented in *Moneyball*, says, "[B]y and large you are what you are" at 27. Einstein, who published his theory of relativity at age 27, might well have agreed.

Despite occasional flashes of zing, this #27 crew is a ragtag collection. "Discipline 27" by Sun Ra, for example—roiling discordance with no discernable meter or melody—sits cheek by jowl with Kool Keith's "27 Shots," on which the rapper/producer takes aim at multiple targets including the corny, the fake-ass, and the apple-haired. As for songs actually titled 27, there's been a resurgence. Recent offerings have tended toward the histrionic, including "27" by Fall Out Boy ("If home is where the heart is/then we're all just fucked"); "27" by the Scottish band Biffy Clyro, whose fans are known to thump chests and shout "Mon the biff!" in solidarity; and "27 Steps," by Georgia pop-rockers Cartel, who downsize the central conceit of "Twenty-Nine Ways (To My Baby's Door)" for leaner times. In "27 Jennifers," former Soul Coughing main man Mike Doughty chronicles the profusion of like-named girls in his high school, a rare exercise in stripped-down, straight-head pop from a guy known for his multilayered hip-hop flavor. Dave Matthews (who also has a song called "27") liked it so much that he eventually released a collection of Doughty's songs, *Haughty Melodic*, (an anagram for the songwriter's name) on his own label.

A SPATE OF 27 SIGHTINGS

— of the 250 soldiers on Iwo Jima hill in Johnny Cash's rueful "The Ballad of Ira Hayes," "only twenty-seven lived to walk back down again."

— Jars of Clay rocked up Psalm 27 and called it "Run in the Night."

— Nelson Mandela spent 27 years in prison (but in 1984, when the Special AKA song "Free Nelson Mandela" was a hit, he had endured "21 years in captivity").

— the prolific protagonist of Bob Dylan's "Rambling, Gambling Willie" had "twenty-seven children, yet he never had a wife."

If 6 Was 9

THE VERDICT

That's a lot of randomness, I know, but 27 has real resonance in these climes. The 27 Club (or, to quote Wendy O'Connor Cobain, "that stupid club") refers to the seeming preponderance of influential musicians who have died at that age, including Janis Joplin, Jimi Hendrix, Brian Jones, Jim Morrison, Kurt Cobain, and Robert Johnson.[1] Amy Winehouse joined the club in 2011. (Her passing, on July 23, would no doubt confirm William Burroughs' theory about 23 and its deathly associations.) A few days later, M.I.A. released a song called "27," previously a demo, which attacks the Rock Martyr mythos over a tribal-meets-space-age rhythm: "All rock stars go to heaven/You said you'll be dead at twenty-seven."

My choice for top #27 song is not so literal. According to Michael Bond, half of the Chicago-area duo Coltrane Motion, "Twenty-Seven" is "a song about writing songs about how you're going to die, and sort of flipping the musician-as-martyr thing a bit." In the first verse, "it takes the 'fame = early death' equation to ludicrous lengths."

Don't clap your hands or I'll die in a plane crash
Don't sing along or I'll overdose
I know you want me/ wrapped in plastic
What a shame, I've got their first album, I loved that song where he went
I'm going to fall apart
in a sea of broken synthesizers
I'm going to walk out while it's still playing
kick drum, snare drum, kick drum, snare drum …

"Twenty Seven" sounds a feedback-drenched note that recalls such initialized U.K. acts as JAMC and MBV, and while you can hear these and other influences, the band assuredly leaves its own thumbprint. Maybe it's because they write their own sound software. In any case, I doff my cap to Coltrane Motion for writing a song that captures 27's tragic ethos through a smooth amalgamation of hooks, feedback, and plenty of kick drum, snare drum, kick drum, snare drum …

A SONG TITLE	AN ALBUM	A BAND	A LYRIC
"Song of 27" – Richard Buckner (1997)	*The Twenty-Seven Points* – The Fall (1995)	27 Devils Joking – New Mexico, punk (1985–1992)	"He was born in the summer of his 27th year" – John Denver, "Rocky Mountain High" (1972)

---

1 In tune with this deathly association is a line from Sufjan Stevens' curiously affecting "John Wayne Gacy": "Twenty-seven people/Even more, they were boys".

28

The menstrual cycle is a pretty universal fact of life on earth, yet it's understandably scarce in the world of song. "28 Days" by LFO[1], for example, isn't about *those* 28 days, just the 28 days "till I see you." On the other hand, "28 Days"—the lead track from the Duncan Guy Band's self-explanatory *Menstrual Album*—is. Also in the demimonde of the human body, officially there are 28 so-called never events: things that absolutely, positively should not occur in a health care setting. The list includes patient abduction, botched artificial insemination, amputations performed on the wrong body part, and other iatrophobic nightmares. But due to the limited musical appeal of "Aunt Flo" and medical mega-mistakes, along with 28-day rehab stints, 28 seems destined to remain a stealthy presence in pop music.

What to make of the agitprop strain running through 28 is anyone's guess, but the numeral loosely joins two songs-as-political-cris de coeur by English outfits that would later find chart success with less edgy works. "28/8/78" is an early offering by Scritti Politti from the *Skank Bloc Bologna* EP, angry, angular stuff with a King Crimsonian precision. Icy voices of British newsreaders swim in and out of the mix, reporting on youth riots and ending with a smugly enunciated " … scores of loudspeakers were pounding out nonstop reggae and rock music." On the one hand, it's hard to believe Scritti's Green Gartside evolved into a purveyor of silken, semiotically sound '80s synth-soul, but then again, the pop landscape is rife with extreme shifts from the outer fringes to the middle of the road.

Ask Chumbawumba, whose 1997 worldwide smash, "Tubthumping," was a hit among the very louts who would not have come within a vomit-arc's distance of the band's previous 12 records, including "Smash Clause 28" from 1988. North of the London climes of Scritti and Chumba, Vini Reilly and the second incarnation of Durutti Column honored an address in his native Manchester with "28 Oldham Street." Reilly's distinctive guitar tone is like umami, its own flavor.

The age of 28 is hardly iconic, yet it is the main subject of the essential #28 canon. The consensus seems to be that if you're 28 and unmarried, unsettled, or still living with your parents, you are in a sad state indeed. As David Foster Wallace remembered at age 34, "I was twenty-eight years old, and that means not taking an advance for stuff before it's done." Folksinger John Craigie's poignant "28" references members of the Forever 27 Club of dead musicians, unable to fulfill the song's delicate refrain, "If I could only make it … to 28." Lorene Scafaria's "28," heard in Drew Barrymore's *Whip It*, is the other side of that coin, a hopeful ditty that proclaims, "When I turn 28/things are gonna be great."

But of course they will.

---

1 As in Lyte Funkie Ones, an American boy band, not British techno pioneers LFO, or Low Frequency Oscillation.

Who knows what future generations will make of the trove of material produced by Hasil Adkins, the self-taught West Virginian songwriter who died in 2005 after being run over by a teen punk riding an ATV. To borrow a helpful phrase from Discogs, "Recurring themes in Adkins' work include love, heartbreak, hunchin', police, death, decapitation, commodity meat, aliens, and chicken." But "Twenty-Eight Years," a country dirge from 1994's *Achy Breaky Ha Ha Ha*, has little in common with typical Adkins fare like "Chicken Walk" and "Sally Woody Wally Waddy Weedy Wally." At 95 percent achy breaky and 5 percent ha ha ha, "Twenty-Eight Years" turns the Adkins MO on its head.

United in 28 and a certain feralness are Mr. Adkins, whose first real release was called *The Wild Man*, and a band whose place in rock history is secure by virtue of "Born To Be Wild." No one would argue that Steppenwolf's "28" is as memorable as that tune, but it's still a pretty choice slab of pumped-up late-'60s rock, with a pounding go-go beat and a laid-back message to offer: "She's 28 years old tonight/I told her not to fear/It's all very right." The song breaks into a bit of mystical gibberish, but in the end "28" is just a tale of lust that ends successfully ("She woke up … grinning with laughter").

And at the age of 28—imagine that.

BUCK-O-NINE — San Diego, ska-punk, "28 Teeth"

Formed 1991, Derivative, Frat-friendly, Member named Andy

SPONGE — Detroit, alt-rock, "Twenty-Eight Days"

THE VERDICT

Toni Basil will forever be considered the Platonic Ideal of the One-Hit Wonder. Now, she did create one of the most aggressively annoying singles ever, but that label does her a disservice. Most one-hit wonders are not nearly as accomplished as the woman born Antonia Christina Basilotta in Philadelphia. Long before she became famous singing "Anyway you want to do it/I'll take it like a man," Ms. Basil had acted with Jack Nicholson in *Five Easy Pieces*, danced with Davy Jones in the Monkees cult favorite *Head*, choreographed a video for Talking Heads.[2] Still, when people think of Toni Basil, they see a grown-ass woman in a cheerleader costume doing stiff-armed poses in the "Mickey" video.

---

2 It was Ms. Basil's fondness for the Monkees' Mickey Dolenz that led to a critical change in her signature song, originally titled "Kitty."

David Klein

Ms. Basil had started working as a dancer and choreographer when she was in her late teens. By 23, she'd earned the right to sing a worldly lament like "I'm 28," which achieves an admirable level of female angst for a song written by a man.[3] "I'm 28" feels spiritually related to Leslie Gore's 1964 anthem, "You Don't Own Me," also in 3/4 time, in which the teenage singer sets the record straight for her meathead boyfriend. One can easily imagine the spitfire from "You Don't Own Me," 10 years later, singing "I'm 28"—older and wiser, yet starting to second-guess this whole taking-no-crap-from-men thing.

When I close my eyes and listen to "I'm 28," I envision the dude from Steppenwolf's "28" chatting up Toni in a roadside bar, telling her, "Hey baby, it's all very right." But Toni's not having it. She tosses her drink in his face and storms out, and as she maneuvers her Corvair home through driving rain, unleashing this song as she goes: "I'm 28, it's getting late/what have I got to do?" she wails. "Lipstick, pancake, shadow for the eyes/it's all been advertised/but it's getting me nowhere."

It's 1966, man. She's tired of the games and the bullshit and the sexism, but the prospect of spinsterhood isn't too attractive either. Yet with a voice and a manner this sassy, it's highly unlikely that Toni will end up "alone and in a rocking chair."[4] In a kinder and wiser alternate universe, Toni Basil would be more famous for "I'm 28" than the rah-rah, mutton-dressed-as-lamb paean to anal that is "Mickey."

What a pity.

A SONG TITLE	AN ALBUM	A BAND	A LYRIC
"28" – Stars of Heaven (1999)	*The Great Twenty-Eight* – Chuck Berry (1982)	28th Day – San Francisco, psych-rock (1980s)	"And I checked my driver's license I was 28 years old" – The Mountain Goats, "Birthday Cake Song"

---

[3] Graham Gouldman, the producer and songwriter behind hits for the Hollies and the Yardbirds and a future member of 10cc.
[4] On the other hand, the protagonist in Little Jackie's "28 Butts," who drowns her sorrows in whiskey and cigarettes and dreams of being a housewife, might well be rocking chair bound. Or worse.

29

FACT: *New Miserable Experience*, the Gin Blossoms' debut, includes a song called "29."
FACT: W.C. Fields, whose gin blossoms inspired the name of this Tempe, Ariz., band, was born on Jan. 29, 1880.
FACT: Tempe, Ariz., which was incorporated on November 29, 1894, has a median age of 29, according to a recent census.
FACT: The early Internet meme known as "Complaint From Seat 29-E"[1] begins, "Dear Continental Airlines, I am disgusted as I write this note to you about the miserable experience I am having sitting in seat 29E on one of your aircrafts."

It's hard to know what to make of the implication that somehow 29 + gin blossoms = miserable experience, but there it is. If this connecting of dots feels a bit forced, my next assertion stands on solid ground: 29 has a powerful connection to the U.S. West Coast. Twenty-nine, after all, is the atomic number of copper, which is mined almost exclusively in the West. Twentynine Palms is a small city in the Mojave Desert that is the home of Joshua Tree National Park, where the Stones took mushrooms and wore knitted ponchos and watched for UFOs with Gram Parsons. The city was commemorated in song by a guy named Allie Wrubel, whose "The Lady From 29 Palms" was a hit for the Andrews Sisters in 1947 and was recorded by folks like Peggy Lee, Doris Day, and Frank Sinatra. Wrubel, an actual resident of Twentynine Palms, also co-wrote "Zip-a-Dee-Doo-Dah," which surely eased his way into the Songwriters Hall of Fame.

Robert Plant's "29 Palms" is nowhere near as 29-centric as Wrubel's "Lady." Yes, a "desert heart" gets invoked, but it's more of a metaphorical desert than one in San Bernardino County that's home to the world's largest Marine base. Straddling the line between geography and metaphor is Bruce Springsteen's "Highway 29," a dark tale of love on the run built around Bruce's spare vocal and that evocative "Streets of Philadelphia" synth wash.

Sublime's "April 29th, 1992 (Miami)" recounts the L.A. riots that followed the acquittal of cops accused of beating Rodney King close to death, and implies that the band members themselves not only took part in the looting but also quite enjoyed the experience.[2] The song ends with a listing of cities, including Miami, Tuscaloosa, and, you guessed it, "fuckin' 29 Palms."

The surprise is that a number like 29 is so well represented across the genres. There's a jazz standard ("29th and Dearborn"); a laid-back jailhouse ditty from country star Johnny Paycheck ("11 Months and 29 Days"); a notably brief, insanely tight two and a

PALATE-CLEANSING #29 SONG:
On the first few releases by the Brooklyn band Goes Cube, songs were numbered rather than named. "Song 29," from *Beckon the Dagger God*, combines the sinew and sludge that remains the band's favored mode of expression.

---

1 In which a disgruntled passenger details the multiple indignities of being seated too close to an in-flight lavatory.
2 Goes without saying that the (L.A.-based) Rembrandts' "April 29" (rhymes with "everything is fine") concerns a different April 29.

half minutes of jazz fusion by drummer nonpareil Billy Cobham ("29"), and a heart-tugging lament for lost gridiron glory in Steve Earle's "No. 29." There's *Nuggets*-grade garage rock from the Sparkles, West Texas boys whose "Hipsville 29 B.C." weds a trippy-sounding title to a down-and-dirty ass-shaker. And there's the bludgeoning, "Twenty-9" from Jon Spencer Blues Explosion's semiofficial first release, *A Reverse Willie Horton*. (Before America became familiar with the Willie Horton who committed rape while on a weekend prison furlough granted by Gov. Michael Dukakis, the only Willie Horton who mattered was the formidable center fielder for the Tigers. In 1965, his breakout year, he hit 29 home runs.)

If delving deep into the 10th prime reveals anything, it's that 29 denotes both a miserable experience and an abundance. Take that lady from 29 Palms—she got 29 Caddys, 29 furs, and 29 fellas ("She's a yip-yip-yippy-eyed doll" to boot). Twenty-nine can also refer to a few. Tom Waits sounds an especially Louie Armstrongish note on "29 Dollars," a sleepy blues whose refrain ("29 dollars and an alligator purse") conjures the Waitsian essence of the lack of plenitude. I'm sure that Tom, who later made reference to "29 gypsies in a Cadillac stoned" in the 2004 song "Metropolitan Glide," would not quibble with the winning song here, a (supersingular) prime example of 29 in the lotta-lotta sense.

THE VERDICT

In "29 Ways," the blues titan Willie Dixon boasts that nothing can keep him from his baby's door, and he supports his claim by ticking off the various ways he has of getting there: "One through the basement/two down the hall/When the going get tough/I got a hole in the wall." While he does not supply precise numerical evidence, we must, and frankly are inclined to take the singer's word for it.[3] But there's no debate about the sturdy appeal of Dixon's song, which puts 29 in the spotlight like no other. Koko Taylor's version, with the longer title "Twenty-Nine Ways (To My Baby's Door)," comes from her self-titled 1969 LP, produced by Dixon himself.[4] Both versions are wonderful. Dixon's is sly and understated, while Taylor shakes the rafters with a voice big enough to match the song's boast. No knock against the original, but I have to give top honors to Koko Taylor for going a step further and singing the living hell out of it.[5]

A SONG TITLE	AN ALBUM	A BAND	A LYRIC
"29" – Lloyd Cole (1987)	*29* – Ryan Adams (2006)	29 Died – Thousand Oaks, Calif., industrial (1990s)	"Now there are 29 skaters on Wollman Rink" – Joni Mitchell, "Song For Sharon" (1976)

LAST WORDS:
The church bell chimed 'til it rang 29 times for each man on the *Edmund Fitzgerald*.

---

3 The same cannot be said for Paul Simon, who almost certainly borrowed the central idea of Dixon's song and flipped it around when he wrote "50 Ways to Leave Your Lover."
4 Another Willie Dixon song—Ms. Taylor's oft-covered signature hit, "Wang Dang Doodle"—is surely the lone point of agreement between Polly Jean Harvey and Ted Nugent.
5 Susan Anton, of *Goldengirl* fame, and Jim Belushi are among the many self-described vocalists who could not resist wrapping their vocal chords around Willie Dixon's classic #29.

30

> "John Henry said to his shaker
> Shaker, why don't you sing?
> I'm throwin' thirty pounds from my hips on down
> just listen to that cold steel ring."
> —"John Henry" (traditional)

Like John Henry, 30 is traditional. The New Testament's 30 pieces of silver has remained a linguistic touchstone and now refers to mercenary treachery in general. The Thirty Tyrants, also known as the Thirty, ruled post-Peloponnesian War Athens with an iron hand. The Roman numeral for 30 lives on as the basis of a slang signifier for a hardcore porno movie (the XXX "rating"). Often enough, though, 30 is just an extremely handy approximation: the "30 seconds time" in which she said "I want to live like common people," for example, in Pulp's outrageously good "Common People," or the 30 guineas Ray Davies has to shell out for "a genuine Hawaii ukulele" in the Kinks' "Holiday in Waikiki."

Thirty is a decade number, so the definitive song has the added obligation of incorporating 30 in a meaningful way. Much as this fact might dismay the goateed, fedora-sporting nighthawks out there pulling for "16 Shells From a Thirty-Ought Six," they can rest assured that Mr. Waits's moment in the numerical sun is coming up (hint: he's riding with Lady Luck). The forecast is not quite as rosy for Humble Pie, who never got much respect from the critics but whose "30 Days in the Hole" was nevertheless a big rockin' hit in 1972. "30 Days," and Humble Pie as a whole, encapsulates a distinct strain of lumbering riff-rock that sounds, in these processed times, refreshingly uncalculated. And the Pie's treble run of genius-level album titles—*Rock On*, *Smokin'*, and *Eat It*—may never be topped.

Thirty's utility as a measurement has led to a toolbox-full of incrementally titled tunes, like Simple Minds' "30 Frames a Second," Spoon's "30 Gallon Tank," and Bow Wow Wow's "C30, C60, C90," an ode to cassette tape formats. Superchunk taps a similarly youthful-exuberant vein in "30 Xtra," a mighty roar from *No Pocky For Kitty*, and "Thirty Whacks" by the Dresden Dolls is a haunting cabaret-style take on the Lizzie Borden story containing this splendid line: "How did I manage to station myself in harm's way/and only get hit with a ticket for loitering that I have no way to pay?"

Our shortest entry, a 14-second goof by the Red Hot Chili Peppers called "Thirty Dirty Birds," reveals just how elusive a New York accent can be for Michigan natives like Anthony Kiedis. When Minnesota native Bob Dylan refers to 30 in "Clothes Line Saga" ("It was January the 30th/and everyone was feelin' fine") his vocal brings to mind a spry but ancient bluesman of indeterminate origin. Of similarly vague provenance are the voices from found answering-machine tapes that course through "Thirty Incoming" by the Books, an English duo whose ramshackle musical collages never lose their deeply felt musicality.

Two top contenders for #30 honors couldn't be further apart. In one corner is a harrowing, hallucinogenic sprawl, and in the other, a Chuck Berry song. It's hard to believe both were spawned in the same universe. "30 Seconds Over Tokyo,"[1] the first single by "avant-garage" avatars Pere Ubu, begins with a few shards of guitar, and already we're in a strange land. A unifying lick emerges, with a tone not far from Tony Iommi's patented sludgy Gibson SG sound. But when David Thomas's damaged tenor enters with a sinister singsong melody, almost childlike, the effect is far more unsettling than your standard satanic imagery. Soon the song's bombing raid commences, which devolves into a tangle of pure sonic chaos. Finally the smoke clears, that guitar lick crawls out, and Thomas and company "take it home."

On a far less fraught side of town is Chuck Berry's "Thirty Days," a hit in 1955, when high school heartache and hijinks were a songwriter's bread and butter. Chuck probably never envisioned a day when disaffected youth, like the members of the Cleveland-based Pere Ubu, would write songs inspired by the whir, whine, and hum of nearby factories. Berry unleashes his tale over Jasper Thomas's scrambling two-beat groove, and it's Berry's hoarse exuberance, the sheer fun in his enunciation of "worldwide hoodoo," the quintessentially Southern pronunciation of "again' ya," that puts it over the top for me. There's also an aggressive solo in the middle that might have inspired Gang of Four's Andy Gill to pick up a guitar. In Chuck's hands, a simple tale of boy loses girl/boy wants girl back, somehow becomes a whimsical riff on paranoia and stalking, with the FBI, the UN, and a gypsy woman all making appearances.

It would be pointless to argue which is better. One is spry and swinging, the other is corroded and menacing. Chuck stirs me on a rhythmic level. Ubu assaults my mind. Both kinds of songs have their hallowed place in my consciousness, but given the trifurcated nature of the numeral in question, something compels me to go for a third option.

THE VERDICT

And as fate would have it, striding over an *El Topo* landscape is a new presence altogether, a "30 Century Man." Scott Walker (born Noel Scott Engel in 1943) is one of pop music's great enigmas. After establishing himself as a hit-maker in England with the Walker Brothers in the late '60s (he even had his own TV series), the increasingly reclusive Walker veered off the rock circuit while continuing to release records sporadically.

Walker has always embraced embellishment—whether the musical theater accompaniment of his early work, Bowie's *Low*-era electronics, easy-listening strings, or Righteous Brothers-style Wall of Sound. But on "30 Century Man" it's just him and his guitar, and that's more than enough. Even within the confines of a surrealistic folk song, Walker's naked voice can't help but resound all the way to the cheap seats. It's only a minute and a half long, but there are details to savor. Grand themes are evoked with nonchalance ("See the dwarfs and see the giants/Which one will you choose to be?"). The chorus suggests a concise escape-

---

[1] "30 Seconds Over Tokyo" takes its name from the title of a book about the famed "Doolittle Raid" of 1942, in which a fleet of American planes bombarded targets in Tokyo, proving the city's vulnerability and helping to turn the tide in the Pacific war. The 1944 film version starred Van Johnson—who played arch-villain The Minstrel on the ultra-campy *Batman* TV series—and Spencer Tracy, who would have made a great guest villain on the show.

to-the-future scenario before the whole thing makes a sudden shift into the mystic as Walker foresees himself shaking hands with Charles De Gaulle. It's one of the hardest lyrical right turns in pop music, right up there with, "That night on our honeymoon/we slept in separate rooms" in Freda Payne's "Band of Gold." Seriously though—Charles De Gaulle? Never saw it coming. Nor the music-box outro.

A SONG TITLE	AN ALBUM	A BAND	A LYRIC
"30" – PJ Harvey (2001)	*30* – The Buzzcocks (2008)	30 Seconds to Mars – L.A., post-grunge/neo-prog ('90s–2010s)	"Son, I'm 30/ I only went with your mother cause she's dirty" – Happy Mondays "Kinky Afro" (1990)

POSTSCRIPT: In *Inventory*, a compendium of lists by the Onion's AV Club, "30 Century Man" is included among "100 Killer Songs Clocking in at Two Minutes or Less." But there is a distinction to be made. As I see it, short songs fall into a few distinct categories: short songs, miniatures, fragments, and instrumentals. A short song is a fully formed song crammed into a small space, like Elvis Costello's "Mystery Dance," or "All Shook Up," by that other Elvis. It has verses, a chorus, a bridge; it's just in a hurry to finish. Instrumentals speak for themselves. Fragments start in the middle, or lop off something critical, like a bridge or an ending.

A miniature, it seems to me, is a brief song that, by virtue of its musical and lyrical particulars, suggests something larger that you can't quite put your finger on. Like a fragment, it might omit certain things, but a miniature feels more fully formed than a fragment, while falling short of full songdom. "30 Century Man" strikes me as a perfect embodiment of this intriguing subgenre (one more: "Final Day" by Young Marble Giants, spare, almost Joycean at "1:43"). At this key decade juncture, it's good to remember what one person playing a guitar and singing can accomplish. Throw in a giant, a dwarf, and some saran wrap, and the possibilities are endless.

LAST WORDS:

"Now lead!"—Ned Plimpton to Captain Steve as "30 Century Man" plays in *The Life Aquatic With Steve Zissou*

31

"And God said: DELETE lines One to Aleph. LOAD. RUN.
And the Universe ceased to exist.
Then he pondered for a few aeons, sighed, and added: ERASE.
It never <u>had</u> existed."

And Arthur C. Clarke submitted the foregoing 31-word short story, "siseneG," to *Analog* magazine in 1984, and some felt the master was being arrogant. And I have just risked incurring the wrath of the Clarke estate by reprinting the work in its entirety. But this is to prove a point. As Clarke demonstrates, it takes a mere 31 words to conjure, and then dispense with, all existence. Yet 31 in the world of song has nothing approaching universe-size ambitions. The numeral suffers, as all numbers ending in 1 do, from having to come after a real showstopper. It's like having to follow Sammy at the Copa.

"The 31st is when I pay the phone bill," quoth the Waitresses in "No Guilt," but seriously, does anything come to mind when you think of 31? The card game also known as Scat or Blitz? The number of ounces in a Starbucks "Trenta" serving? I know, that's reaching. But it's no stretch to say that 31 is the vital nexus between baseball and organs. Let me explain. Denny McLain, who won 31 games in 1968, is the last man to win more than 30 games in a season (and very likely the last who ever will). McLain's career was prematurely derailed by his gambling mania, but during his few years as one of the dominant pitchers in baseball, Capitol Records actually released two LPs by the Hammond X-77-playing hurler. *At the Organ* includes McLain's enervating take on Donovan's "Hurdy Gurdy Man." Seek it out if you dare.

Far more edifying in the 31-baseball-organ genre is the wonderfully titled "31st Season in the Minor Leagues" by Magnolia Electric Co., an alt-country ode to failure in which the tremulous thrum of a Wurlitzer supplies a palpable ache. In "31 Candles" by the Mendoza Line (named for shortstop Mario Mendoza, whose .215 lifetime batting average has come to signify that elusive benchmark known as "just good enough") the organ in question is the johnson of singer Shannon McArdle's soon-to-be-ex, who happens to be her bandmate, Timothy Bracy. (McArdle mockingly suggests that his latest harpy is building a shrine around it.) "31 Candles" deserves to be enshrined in Kiss-Off Song Heaven.

Thirty-one might be best known for its role in a dreary yet useful little rhyme that we learned in school: "Thirty days hath September/April, June and November/All the rest have thirty-one/Excepting February alone …" (Some rhyme, eh? A white person definitely wrote that.) My associations with the poem are uniformly negative because I tend to recite it whenever I have to commit to something. Still, 31 has been pretty damn kind to the Baskin-Robbins people (we'll get to that), and Nick Hornby's *31 Songs* (it was called *Songbook* in the States) is one great piece of music writing. Even the Beatles got around to making a reference to

31, Paul McCartney did anyway, in a song John Lennon famously detested, "Maxwell's Silver Hammer." I always heard it as "He's seen 31," but the line is "PC 31 says he caught a dirty one." P.C. stands for police constable, but it sounds more along the lines of Bowie's "TVC 15," which, it was said at the time, had something to do with masturbation. Which brings us back to 31, Turkish slang for masturbation ("otuz bir"), according to reputable online sources.

I plead guilty to charges of grasping at straws. But slim the pickings may be, I submit they are meaty. "Pluto, September 31st" by the Moving Sidewalks is an early stomping ground for Billy Gibbons of ZZ Top. "Pluto" is the epic-length track from the Sidewalks' sole album, *Flash*. Gibbons does his best Hendrix impression, and drummer Dan Mitchell's Mitch Mitchell impression might be even better. But Hendrix's lyrics never descended into *Shark Sandwich*-era Spinal Tap like, "A mystic fog is in my eyes/ the carpet's been pulled from under my butt/and as the dark begins to clear/my brain's reduced to one watt."

That's actually a pretty fair description of how I feel listening to "Thirty-One," by West Virginia's Karma to Burn, a band that doesn't muck around with names of songs, or even with words. They just play hard and fierce. Like the other offerings on *Wild Wonderful Purgatory* ("Twenty," "Twenty Two," "Three"), "Thirty One" is a multipart headbanger that would please both Dave Mustaine and Butthead. All hail not giving a shit about song names. That's just ballsy. *You want a title? Fuck that. Next song! What are we up to? 32? Thirty-two! Keeerraaang!!!"*

Ellen: "How could you possibly be in love with Miss Fingerwood?"
Pete: "I don't know. She swam across the Mississippi. On Christmas."
Ellen: "She's 31 years old and smells like chalk."
—*The Adventures of Pete and Pete*, "The Valentine's Day Massacre"

A good 31st-birthday song is hard to find, but best not to start off a 31st-birthday mix with Aimee Mann's "31 Today." Lead up to it, or slip it in the end. It's not where you want to begin. Mann likes shining a light on moments that most people prefer to keep hidden, like the soul-stealing awfulness of a Sunday high, or sulking inside on the Fourth of July, resenting revelers. Here, turning a year older begins with a guilty afternoon beer, progresses to uninspired sex, and ends with "getting loaded and watching CNN." It hardly matters, though. Aimee's warm chords and sweet-and-sour vocals are as beguiling as a guilty Guinness on your birthday.

THE VERDICT

Once again, I find myself tipping my cap to the Shirelles, four high school friends who started as the Poquellos in 1958. Brassy and sassy and tuneful, the Shirelles were capable of musical coquettishness, along with what music writer/DJ Charlie Gillett termed "vulnerable and suppliant availability." Their vocal sound echoes through the Motown era to the lady soul singers of the '70s and onward. While "31 Flavors," released as a single in 1963, may not carry the emotional weight of "Baby It's You" or "Dedicated to the One I Love," as infectious trifles go it's hard to beat. I am putty in its hands from the trilled syllables that herald its arrival: "*Ahhhh, yah yah yah yaaaah …*"

David Klein

One of the more effortless laughs in the generally apoplectic comedy *It's a Mad, Mad, Mad, Mad World* is the initial appearance of Sylvester, the beatnik son of Ethel Merman,[1] who manically gyrates to an alternate, much faster version of "31 Flavors." As in so many popular songs of this era, the sexual subtext is buried under a lot of ice cream, but below the froth lurk plenty of things that lead to close dancing. Obviously the term "31 flavors" is not original; Baskin-Robbins came up with that, in 1953, to distinguish itself from Ho-Jo's, which was proud of having 28. But there's a certain rightness to 31 flavors; it's an amount that suggests a multiplicity of colors, textures, and tastes, but not so many as to completely boggle the mind with choices. And a love object named Ice Cream Joe who can kiss in 31 flavors suits the notion of kissing just right, in a way that, say, Belle & Sebastian's "String Bean Jean" never could.

A SONG TITLE	AN ALBUM	A BAND	A LYRIC
"Highway 31" – Champion Jack Dupree (1946)	*Gimme Samoa: 31 Garbage Pit Hits* – Angry Samoans (1987)	The 31st of February – Jacksonville, Fla., rock (1960s–1970s)	"Oh, take me to the station/ cause I'm number 31" – Creedence Clearwater Revival, "Graveyard Train" (1968)

POSTSCRIPT: In Snoop Dogg's "31 Flavors," he makes it clear that no one messes with his ice cream truck.

---

[1] Played by Dick Shawn, who also tore up a beatnik-y role as Lorenzo St. Dubois/Adolf Hitler in the original movie version of *The Producers*.

If 6 Was 9

	1 "First of May" –The Bee Gees	**2** "Day After Tomorrow" – Tom Waits	**3** "July 3rd" – People Under the Stairs	**4** "May 4, 1951" – Candidate	**5** "Fifth of July" – Terry Reid	
6 "Damsel of Death (Sixth of June) – It Dies Today	**7** "February 7" – The Avett Brothers	**8** "April 8th" – Neutral Milk Hotel	**9** "March 9th" – Busta Rhymes	**10** "June Tenth Jamboree" – Louis Jordan	**11** "Love and War 11/11/46" – Rilo Kelly	**12** "July 12, 1939" – Charlie Rich
13 "September 13" – Deodato	**14** "Avril 14th" – Aphex Twin	**15** "Happy Birthday to Me (Feb. 15th)" – Bright Eyes	**16** "June 16th" – Minutemen	**17** "December 17" – Jean Michel Jarre	**18** "June 18, 1976" – Pedro the Lion	**19** "(March 19th 1983) It Was Probably Green" – Carissa's Weird
20 "4-20-02" – Pearl Jam	**21** "The 21st" – Blue October	**22** "January Twenty Something" – Why?	**23** "12/23/43" – Poison the Well	**24** "Jan. 24" – Lambchop	**25** "25th December" – Everything But the Girl	**26** "October 26" – The Pretty Things
27 "Kentucky, February 27, 1971" – Tom T. Hall	**28** "28/8/78" – Scritti Politti	**29** "Someday (August 29, 1968)" – Chicago	**30** "The Last Day of June 1934" – Al Stewart	**31** "Pluto, September 31st" – The Moving Sidewalks	"32nd December" – Babyshambles	

32

Water freezes at 32 degrees Fahrenheit. There are 32 teeth in the healthy human mouth, and the .32-caliber revolver—the 32 gun Bad Bad Leroy Brown kept in his pocket for fun—has left its mark. So it's surprising that the pickings are as select for 32, an elegant power of 2, as they are for the obstinate prime 31. When 32 does appear in song, it rarely seems essential to the proceedings. But, to quote "Oedipus" (not the play, the Regina Spektor song) "32 is still a goddamn number" and must be given its due.

Why Peter Gabriel went with 32 in "The Chamber of 32 Doors" is known only to Peter and the robed art-rock gods on high. The track comes from *The Lamb Lies Down on Broadway*, a four-sided concept LP by Genesis about the spiritual journey of a graffiti-scribbling Puerto Rican youth named Rael. The record's urban setting has always confused me, because all Genesis music emanates from a fog-enshrouded glade in Avalon, right? How could Gabriel, reedily intoning lines like "I'd rather trust a country man than a town man" amid a sea of treated keyboards, ever evoke Times Square? They Might Be Giants, on the other hand, are strictly urban; they couldn't evoke a glade if they (might be) tried. These days they feast on the kiddie market, but the brief, twitchy "32 Footsteps" typifies TMBG's early eccentricity.

Being 32 years old hasn't often been treated as a subject worth singing about. In "Right Here," a gorgeous if disconcerting love song by the Go-Betweens, Grant McLennan sings, "I know you're 32 but you look 55." It would make a fitting segue from Aimee Mann's "31 Today" on that depressing birthday mix. About the only major figure to weigh in on how it feels to be 32 is Ringo Starr, who minces no words in "I'm the Greatest," proclaiming, "Now I'm only 32/and all I wanna do/is boogaloo." No one would write a line like that anymore. But when John Lennon dashed it off in a song he later gave to his former drummer, "boogaloo" was a perfectly acceptable, even hip way of saying get your freak on.

If you're unfamiliar with Van Morrison's "Thirty Two," that's probably a good thing. In 1967, he somehow "made good" on his contract with Bang Records by cranking out a bunch of spoken-word crapola as devoid of merit as anything ever produced by someone of Van the Man's stature. Not car-crash fascinating, though. Just bad.

Similarly spoken-word but not at all heinous is the poet Kenneth Rexroth's "State & 32nd," which nevertheless supports the notion that in general poets should avoid fronting bands, unless they're Patti Smith. It's actually not bad. Rexroth's voice, aptly described as "crabby" in a review I read, has a sly way with lines like, "Dice girls going home/Whores eating chop suey/Pimps eat chile mac/Drowsy flatfeet/ham and eggs." And his sidemen are first-rate, if a bit short on hooks.

Robert Johnson's "32-20 Blues" gets its name from a .32-20 rifle cartridge and its potential to cause harm to an unfaithful woman. The song was released as a single in 1937, a year before the man faded into legend. It's a primal and spare blues, just Johnson playing his guitar and singing ten three-line verses. The narrative alternates between dire warnings and gun talk, specifically

Johnson's preference for the 32-20 pistol over his baby's .38 Special ("I believe it's most too light"). The tempo picks up slightly toward the end, and his couple of "hey heys" in the "where did you sleep last night" verse are almost jaunty, but the song ends on a menacing note: "Ah boys, I just can't take my rest/ with this 32-20 laying up and down my breast."

THE VERDICT

Robert Johnson's greatness is a point of rare unanimity among thoughtful listeners and musicians alike. He is a historic yet elusive figure, and any of his 40 or so surviving recordings has more "value" than just about any pop song you'd care to mention. For me, though, the #32 slot has always belonged to The House of Love, a London band who benefited greatly from luxuries Johnson never had, including modern recording techniques, a cushy record deal, and advances in the art of barbering. After I discovered them while visiting England in 1989, their second eponymous release (aka "the butterfly album") lit up my bright yellow "Sports" Walkman like few others. Those guitars—a perfectly calibrated mix of fuzzy and sweet, wailing and warm—and the dark, anguished vocals of Guy Chadwick became the moody soundtrack to my late 20s. Even the name Guy Chadwick was something I considered surpassingly cool.

"32nd Floor" begins with silence and then, like a dropped match igniting the drapes, roars to glowing life. The ringing guitar lines of Terry Bickers create a masterful interplay between acoustic and glinting electric, circling each other in a slinky two-step as a chin-bobbing beat takes hold and Chadwick's burnished baritone comes in. The words might seem heavy-handed on their own, but in the song they are just right (in a *Ben Hur* kind of way). You don't even mind that "mind" rhymes with "mind."

The scribes tell us The House of Love amounted to a flash of chiming guitars never made good on before the twin waves of acid house and grunge tsunami'd the musical landscape. Maybe so, but guitars that flash this brilliantly will always paint purple-hued visions in my mind's eye, set my chin bobbing, and summon a grin of pure approval.

And I'll never ask for more.

A SONG TITLE	AN ALBUM	A BAND	A LYRIC
"32 Weeks" – The Mekons (1980)	*32 Minutes and 17 Seconds* – Cliff Richard (1962)	Wretch 32 – London, rap (2010s)	"Thirty-two teeth in a jawbone" – Grateful Dead, "Alabama Getaway" (1980)

33

"In 1977 the ceiling on the age for rock musicians wasn't 70 years old, as it is today. It was about 33."
—Bruce Springsteen, *The New York Times*, Nov. 7, 2010

Thirty-three was once the ceiling, says the Boss, and I'm inclined to believe him. When Chrissie Hynde of the Pretenders found herself a mother in her 30s and still leading a professional rock 'n' roll band, she felt compelled to comment on it ("I'm not the cat I used to be/I've got a kid I'm 33, baby!"). Jesus Himself accomplished all his earthly works by the age of 33. But the charisma and self-possession of a Hynde or a Springsteen or an Of Nazareth are rare commodities. Being a successful 33-year-old rock star, never mind rising from the dead, is beyond the reach of most of us.

My own personal ambitions were far more modest back then. At roughly 33, I felt I had high-fived the cultural zeitgeist by managing to have my letter praising James Iha of Smashing Pumpkins printed in *Spin* magazine. My passion for the band dimmed a few months later, though, after I witnessed one of Billy Corgan's infamous anti-crowd rants, and it waned even further with *Mellon Collie and the Infinite Sadness*. As anyone familiar with reading music criticism in the bathroom will tell you, most double albums are better off as killer one-disc releases, but try telling that to a simpering genius like Corgan.

"Thirty Three" exemplifies how, by *Mellon Collie*'s second disc (dubbed "Twilight to Starlight"), Corgan becomes insufferable. Beginning faux-contrite ("I forgive everyone"), he returns swiftly to martyr mode, proclaiming that the earth laughs at "the blasphemy in my jangly walk." I suppose he's saying it's a crime against nature to have spring in your step at the decrepit age of 33, but hey, the earth doesn't know you've got *Fables of the Reconstruction* playing through your earbuds. It just knows you have a jangly walk. And to the earth that's very funny.

Kris Kristofferson's "The Pilgrim, Chapter 33" is one of his most beloved creations, but early on in *Taxi Driver*, Travis Bickle doesn't take it well when Cybill Shepherd's Betsy quotes the song to him ("He's a prophet and a pusher, partly truth, partly fiction/A walking contradiction"). "I'm no pusher," Travis says. "I never have pushed." Betsy tells him, "No, no. Just the part about the contradictions. You are *that*."

The same might well be said of Sinead O'Connor, whose angelic pipes have always seemed at odds with her combative persona. In "33," from *Theology*, her contrary streak remains in evidence as she spices up Psalm 33 of the Bible with an exhortation to crank up one's bass amp for Jah.

A MOMENT OF SILENCE: In John Cage's most famous composition, "4:33," instrumentalists are instructed to refrain from playing during the work's three movements, which last 30 seconds, 2:23, and 1:40, respectively. Audiences in 1952 took it as a provocation, a joke at their expense. But Cage was adamant. "They missed the point," he said. "There's no such thing as silence."

The only time that a somewhat average American teenager might have heard the Verlaines, a New Zealand group led by musician/scholar Graeme Downes, would have been in 1993, when the *No Alternative* collection of B-sides and rarities was issued to benefit AIDS charities. The compilation included songs by Nirvana, Smashing Pumpkins, Matthew Sweet, and two by the Verlaines. From the unresolved opening chords of "Heavy 33," the Verlaines are in pensive territory, with Downes sounding like he's singing from the ledge of a tall building. By the end, the dirge has gained force and strength, despair turning into sullen resolve. Following this release, the band went on a lengthy hiatus.

THE VERDICT

Not many people have covered Stereolab songs. Perhaps the distinct vocal interplay between Laetitia Sadier and the late Mary Hansen simply doesn't lend itself to easy reinterpretation or reenactment. Enter Samuel Beam, aka, Iron & Wine, whose early, determinedly lo-fi acoustic style seemed to render everything he touched into an Iron & Wine song. His take on Stereolab's "Peng 33" succeeds in doing just that. I didn't think it was possible to hear a Stereolab song without those inimitable vocals, but somehow, in a feat of musical alchemy, Mr. Beam has excised the Can and left a campfire in its place. Perhaps he was drawn to the simplicity of the words, which are a long way from the Marxist platitudes of early Stereolab records and more like straightforward expressions of optimism: "Curiosity was far greater than our fear/It felt so simple and so prodigious at the same time/Incredible things are happening in the world."

We should be grateful to Stereolab for introducing a slew of previously uncool styles into the indie pop mix and for singing nakedly optimistic words like the ones above (and getting away with it). Beam's beatific whisper also proves an alluring mode of transport. He takes the quirky essence of a stylized piece of music and transforms it into something timeless. Ultimately, though, Stereolab's enchanting singsong is a siren's call I cannot resist. After no small amount of deliberation, I confer top honors on the 'Lab, with Beam's brilliant reimagining close behind.

A SONG TITLE	AN ALBUM	A BAND	A LYRIC
"No. 33" – The Clientele (2010)	*33rd Revolution* – The Quest (1967)	33 – Spokane, Wash., hard rock (2000s–2010s)	"Hail hail, 33/ Your violent hiss sounds so sweet to me" – Shovels and Rope, "Hail Hail" (2012)

BOTTOM OF THE 33RD is Dan Barry's account of baseball's longest game, a contest between the Pawtucket Red Sox and the Rochester Red Wings, which commenced at dusk on April 18, 1981, and was finally suspended the following morning with the score still tied.

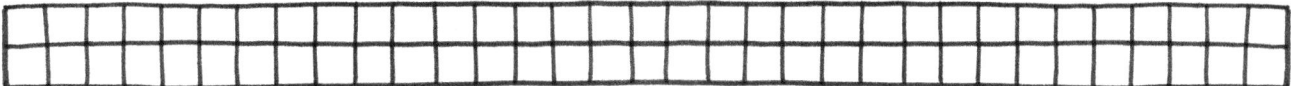

34

Next to 33, a number with style and grace and RPM, 34 is a nerdy older brother named Eugene who doesn't get out much. Think of it: 33 signifies coolness itself—LP records—while 34 (a heptagonal number) represents, well, heptagons. No wonder 33 gets all the glory. There just isn't a lot of thirty-fouria out there. Oh, there's *Miracle on 34th Street*, a mostly heartwarming 1947 Christmas bauble featuring a young Natalie Wood. There's Rule 34 of the Anonymous group's Rules of the Internet ("There is porn of it. No exceptions"[1]), but when 34s appear in song, you find them tucked away between parentheses or linked to an ordinal. Thus the intrepid #34 searcher must sniff them out like truffles.[2]

Sometimes they hide in plain sight, like "The Ballad of Danny Bailey (1909–34)" by Elton John. It's been in my midst for so many years that finding it is like discovering a long-lost cousin living behind the wall. Don't get me wrong: No one has ever referred to track 2, side 3, of *Goodbye Yellow Brick Road* by its proper name. It was simply "Danny Bailey." But it's a #34 song nonetheless.

Elton touched down in this rustic American territory a few times throughout his career, most wonderfully on *Tumbleweed Connection*, so the saloon piano and larger-than-life subject matter are familiar ground. But tuning in to "Danny Bailey" is to feel truly roused by Elton's chordal brilliance, particularly on the sumptuous turn to C major on "Dillinger's dead/I guess the cops won again" to set up a return chorus. The crook-as-antihero mythmaking does tend to date "Danny" to an early-'70s moment wherein the enlightened rogues of the counterculture, like the stars of *Bonnie and Clyde*, were still in vogue, but goddamn, it's still stirring.

"Voyage 34" is a four-phase suite by Porcupine Tree, the brainchild of English multi-instrumentalist Steven Wilson. It simulates the ill-fated 34th acid trip taken by a young man named Brian, and it has all the earmarks of an epic—not just length but also moments of transcendent weirdness in its various crisis points and slam-bang ending. First Brian and his fellow travelers gobble their sugar cubes to the strains of something akin to an acid-house remix of "Another Brick in the Wall." The Floydian flair continues to unspool as ghostly voices fade in and out among sheets of wailing guitar. Still, using "Voyage 34" as the soundtrack for a real LSD trip would be sheer folly. The opening narration alone[3] would send any would-be astral traveler with half a brain left scurrying for Roxy Music's *Avalon*.

Naming an instrumental composition would seem to be one of the simpler tasks a songwriter faces. If you happen to find yourself wordlessly strumming your acoustic in Room 34 of a cheap hotel, you as good as have your title. It's in this humble instrumental category that many #34 truffle-songs lurk, like "M 34" by the film score maestro Ennio Morricone from *Spaghetti Westerns Vol. 1*; and Dave Matthews's gently picked "#34," the ideal musical accompaniment to a scene in which the title character of *House* sits at home popping Vicodin and working out a chess problem. Master jazz pianist Bill Evans and his Trio riffed on the

---

1 Rule 35 says, "If no porn is found at the moment, it will be made."
2 One such truffle of a non-music-related variety appeared in a February 2012 New York Times review of *The Lifespan of a Fact*, a book-length argument between a fearless fact checker and a solipsistic dissembler whose book the fact checker is fact-checking. A rift arises when the writer is asked to explain why he changed the number of licensed strip clubs in Las Vegas from 31 to 34. The writer replies, "Because the rhythm of '34' works better in that sentence than the rhythm of '31.'" Ah, the rhythm of 34. Is he serious? I'm not hearing it.
3 It begins: "This remarkable, sometimes incoherent, transcript illustrates a phantasmagoria of fear, terror, grief, exultation and, finally, breakdown."

numeral's skedaddle implications in "34 Skidoo," while modern jazz guitar master Nick Colionne's "34" pays homage to Walter Payton of the Chicago Bears, who wore that number to pigskin glory.[4]

THE VERDICT

Bernie Taupin's lyrics and the song's very title imbue "Danny Bailey" with a compelling pseudo-authenticity that would be more than enough here—if not for the presence of Charley Patton, a true-life mythic figure who gave up the ghost for real in 1934. Patton, a corn liquor-swilling, woman-chasing brawler, was known as the Father of the Delta Blues. He had a from-the-gut bellow that could be heard for hundreds of yards without amplification. He'd play the guitar on his knees or toss it in the air and catch it between his legs, long before anyone dreamed of mistreating a guitar in such a manner. Often he used his instrument as a drum, simply pounding on it for protracted stretches to keep the beat going during the wild Saturday night dance parties where he played. From all accounts he was as cantankerous as the American military hero with whom he shared a surname.

When the commercial potential of blues first became apparent to the music industry in the 1920s, it was the more homogenized, sophisticated variety that was brought to the cities and nurtured there. Players in the Deep South, like Patton, and their raw, open-ended, rhythmically complex style were not in favor. Only in the last year of his life, when Patton was brought to New York to record his songs, did he receive the recognition of the music establishment. No tapes of the sessions remain; his entire recorded output has been transferred from scratched and abused 78s. These 57 songs are like living fossils, ancient and yet alive with passion and sorrow.

In moments of weakness, Patton would repent and swear to follow the Bible, but soon he would return to his lowdown ways. Toward the end he began to yearn for a bit of stability. "34 Blues" is a lament for a particularly cruel year. His money's gone, his car gets taken away, and it's back to pushing a plow on a farm. Patton's lyrics are difficult to decipher, but his voice's scarred timbre and haunting resonance make the pain he sings of easily felt. Charley Patton never saw '35, which makes "34 Blues" all the more poignant. He sounds like a man without a lot of time left.

"I ain't gonna tell nobody, '34 have done for me
I ain't gonna tell nobody what '34 have done for me
Took my roller, I was broke as I could be."

A SONG TITLE	AN ALBUM	A BAND	A LYRIC
"Island 34" – Candidate (2002)	*34 Number Ones* – Alan Jackson (2010)	34Bliss – Detroit, hard rock (2000s–2010s)	"I was just 34 years old/ and I was still wandering in a haze" – Pete Townshend, "Slit Skirts" (1982)

---

[4] Further on the Peyton tip: "Peyton on Patton," by the Reverend Peyton's Big Damn Band, is a tribute to Charley Patton, who just happens to be the king of this #34 slot. That's kind of wild.

35

When Sinatra croons, "It Was a Very Good Year" (in D-minor, the saddest of all keys), he slips into this twilight-of-a-ladies-man narrative as if it were a smoking jacket and slippers. Fondly recalling a lifetime of dalliances, the aging raconteur pauses to savor Year No. 35, a time of "blue-blooded girls of independent means." Out of context, the phrase has a distinct Westminster Kennel Club feel to it, but in 1961 he was paying these gals a compliment. Elvis Costello's "And in Every Home" summons a different type of 35-year-old ("She's only 35/going on 17"), while the avenger who Andy Pratt sings about in "Avenging Annie" is now "almost thirty-five, and she's found her peace and she's found her release and she's happy just to be alive."[1]

Thirty-five tends to turn up in ordinary circumstances. You see it as a clock-face demarcation ("8:35 on the Dot" by Peter Lee Stirling[2]); a tidy sum (the Aluminum Group's stately "$35"); and as a location signifier, as in Phil Ochs's "Firehouse Thirty Five," where "in between all the fires" this randy troupe is busy "quenchin' their desires." Then there's Penthouse 35, where Neil Young says he'd be content to kick the bucket (as long as he can "Sail Away"). Devoted Stones fans will recall 7:35 as the time when "they queue up for the bathroom" in the diabolically good "Live With Me" from *Beggar's Banquet*. "4:35 in the Morning" is a time of deep pining for Saint Etienne's Sarah Cracknell, while in Aesop Rock's "11:35," the title time is when "some shit went down." Although after several listens I still can't say exactly what that shit was.

"35 in the Shade" brings A.C. Newman's fine first solo collection, *The Slow Wonder*, to a surging finish. The swell of piano chords that courses through the song deliberately recalls the Spectorian ur-pop of "Da Doo Run Run," but here the classic '60's-grade pop paradigm feels deconstructed and put to new uses. While the often-inscrutable Newman leaves us to ponder the title phrase, the emphatic harmony vocals on "*There goes my ride*" shed light on how the card game depicted in the song turned out. Also lovely is the minimalist "Bit 35," from Broadcast's *Tender Buttons*. Sadly, it's just a bit too slight to win out over the lead track of Dylan's *Blonde on Blonde*.

THE VERDICT

"Rainy Day Women #12 & 35" came into view when I was considering #12 songs, but a wise number song maven knows when to hold 'em and when to fold 'em. This may not be my favorite Dylan track, but it's an important and at the time scandalous song

---

1   Never heard of Andy Pratt? *Rolling Stone* once declared that this scion of the oilman who founded the Pratt Institute had "forever changed the face of rock." But somehow the world did not take notice. He remains known primarily for "Avenging Annie," which was based on Woody Guthrie's "Pretty Boy Floyd" and still sounds great.
2   As Daniel Boone, Stirling had a hit in 1972 with "Beautiful Sunday."

from one of his essential LPs. In 1966, a single by the biggest singer in the world with the refrain of "Everybody must get stoned!" was just the kind of thing to give broadcasters fits. Though nowhere near as druggy as "Eight Miles High," "Rainy Day Women" no doubt kicked off many a pot party. Incidentally, it's the only Dylan song to employ a brass band and stands out as one of his rare overtly comedic songs.

While the refrain might suggest marijuana, the stumbling tempo, ramshackle instrumentation, and impromptu hoots and catcalls are more indicative of a drunken barroom sing-along than a bacchanalian smoke-out. The verses certainly don't paint a stoned scenario; they allude to an inevitability—but of what? Being found out? Misunderstood? Screwed over? *Killed*? After all, "getting stoned" is an ancient, torturous death enacted by a hostile community; one can surely hear the song's assurances as reflecting Dylan's paranoia and disgust with an increasingly intrusive strain of hero worship to which he was being subjected. And who is the "they" he keeps referring to? Top candidates include The Man and the crazy fans sifting through his trash bins. But you can never be sure.[3] As for Dylan himself, he would only say this: "I never have and never will write a drug song."

A SONG TITLE	AN ALBUM	A BAND	A LYRIC
"35 Millimeter Dreams" – Garland Jeffreys (1977)	*35 Reasons Why Christmas Can Be Fun* – Spike Jones (1958)	McDonogh 35 Band – New Orleans, marching band (2000s–2010s)	"I remember the thirty-five sweet goodbyes" – Steely Dan, "My Old School" (1973)

---

3 Adherents to the marijuana theory point to the numbers in the title—12 and 35—which when multiplied equal 420, a certifiable slangy reference to pot use, but Dylan's song predates this usage by almost a decade.

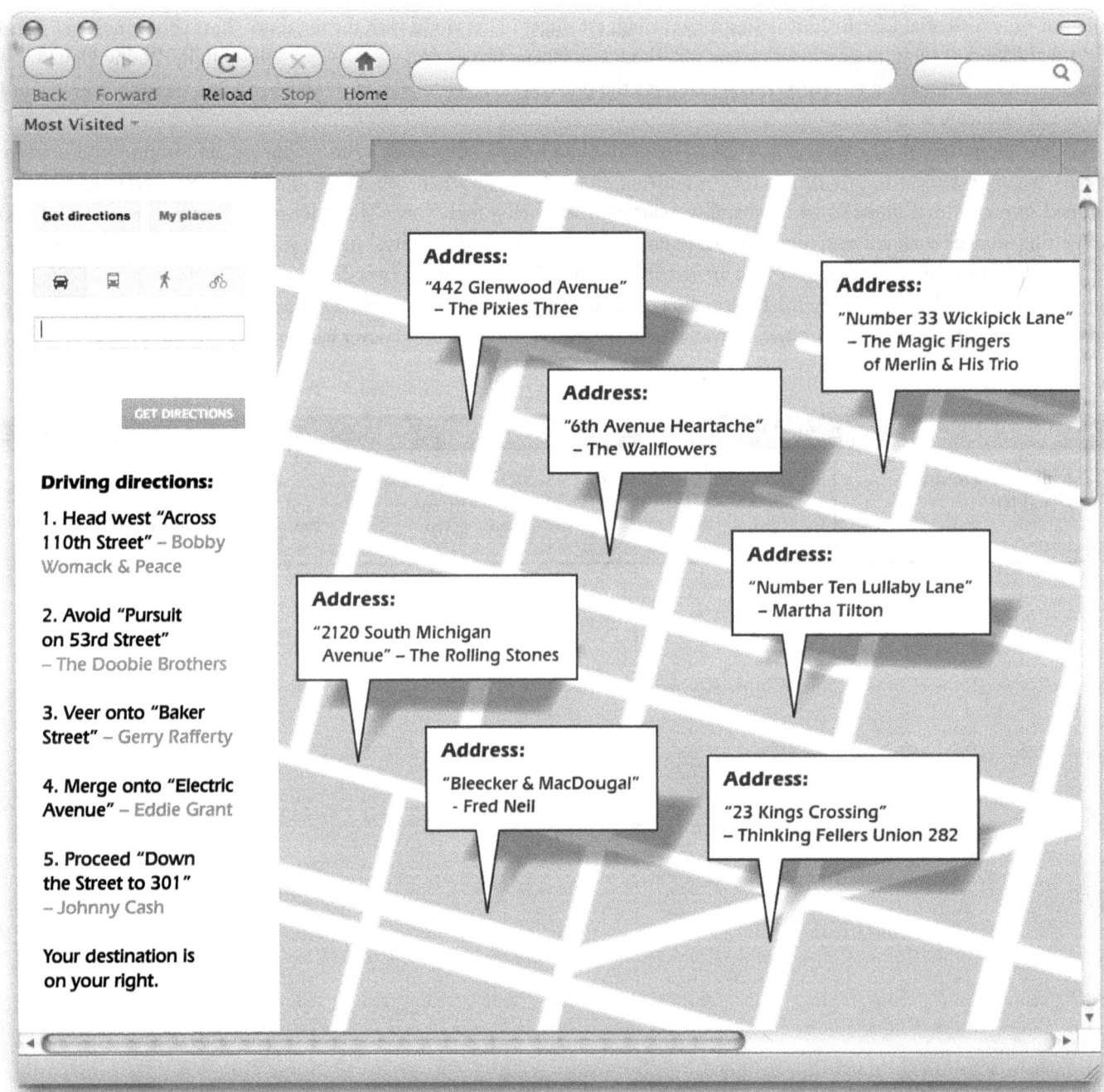

36

Thirty-six is the model figure when it comes to the measurement of women's bust size. No other numeral has such an association with brassieres or has spawned such a bevy of songs extolling female comeliness in such admiring yet quantitative terms. When Walter "Clyde" Orange—the Commodores' lesser-known lead vocalist next to Lionel Richie—lets loose with "36-24-36, what a winning hand!" in "Brick House," he's expressing a time-tested sentiment. Sir Mix-A-Lot might not agree ("36-24-36? Haha, only if she's 5'3""), but he is in the minority. To wit:

SONG	BAND	KEY LYRIC
"36-24-36"	The Shadows	[Instrumental]
"36-24-36"	Violent Femmes	"I want lots of pretty chicks"
"36-22-36"	Bobby "Blue" Bland	"My girl can be tied/ but she can't be beat"
"36-22-36"	Z.Z. Top	"My thing is a real fine thing"
"36D"	The Beautiful South	"So what, is that all you've got?"

A music writer I respect once told me that he never rejects a band on the basis of the singer's voice. I remind myself of Grayson's credo whenever I catch myself wincing at a particularly abrasive set of pipes, but it's safe to say I won't ever get the hang of not minding the voice of Brian Molko of Placebo. OK, I admit, "36 Degrees" is pretty fierce. But for me, the over-serious way he delivers lines like "Shoulders toes and knees/I'm 36 degrees" takes all the fun out of doing deep-knee bends in unison with the song in the first place. Similarly inappropriate for vigorous physical exertion is "Dia 36," by Os Mutantes, who have been called the Beatles of Brazil. Seen in that light, "Dia 36," from 1969's *Mutantes*, demonstrates how even the Beatles' most negligible efforts—dreary pap like "Blue Jay Way"—had a global influence.

Closer to home, the emphatic "Thirty-Six" by Brooklyn's Man in Gray makes me want to smash stuff up. Something about vocalist Christina Da Costa's seen-it-all snarl recalls the great, unheralded Columbus, Ohio, trio Scrawl. But why "Thirty-Six"? I had to know, so I contacted the band and heard back from Jared Friedman and Bryan Bruchman, who provided insight into the circuitous route writers often take in naming their work.

Friedman: "The song was pretty much written, but nobody could think of a name, and someone (I think it was me, but am not 100% sure) jokingly suggested 'thirty-six,' i.e., a totally random title. Pure silly snark. Though as time went by, we came up with all sorts of plausible 'meanings' for it."

Bryan Bruchman (lyricist): "The song itself is about relationships, jealousy, and dealing with balancing your current partner with a previous one who plays an active role in your life (for example, at work, a roommate, in a band). In that sense, the title refers to the movie *Clerks* (which is set and was filmed right near where I grew up), in which the main character [Dante Hicks] finds out that his girlfriend has been a bit more promiscuous than he'd thought, having given blowjobs to 36 guys before she met him."[1]

The poetic narratives of John Cooper Clarke are firmly planted in the muck of his native England, and on 1978's *Snap, Crackle and Bop*, he gives us "Thirty-Six Hours." The song could pass for punk-tinged rockabilly à la Rockpile were its subject matter not a torture session. But in his inimitable delivery, always described in the press as nasal and Mancunian, Clarke still finds some levity in a situation where "everybody looks like Ernest Borgnine."

1966 may not have been a great year for Ernest Borgnine, whose *McHale's Navy* had come down with that crippling show-biz ailment known as "low ratings and repetitive story lines," but for Jim Ford, who went on to make one album that achieved low ratings but had not one repetitive story line, the year was pivotal. It was then that the native of hardscrabble Kentucky ditched academia and landed in San Francisco, with the varicolored songwriting chops he'd developed during his itinerant youth in tow. He soon caught the notice of an enterprising record company owner.

Over the next decade, Ford worked as a songwriter, had his work recorded by some greats and several very-very-goods (Aretha Franklin, Bobby Womack, The Ventures, Ron Wood, Dave Edmunds), and spent a lot of time hanging out with Sly Stone, who once called Ford "the baddest white man on the planet." But the music Ford released under his own name made nary a ripple in the marketplace. His one record, 1969's *Harlan County*, is a longtime cult item that was re-released in 2007 under the title *The Sounds of Our Time*. The renewed interest pointed to a possible comeback of sorts, but Ford, who had grappled with addiction, died later that year in the trailer in Mendocino County where he lived.

THE VERDICT

"36 Inches High" is an odd song even for Jim Ford, who wrote a lot of odd songs. Most of his work draws on New Orleans and Kentucky musical traditions, but this is a whatsit of no clear provenance. He sings it in a plangent tenor, almost a cappella, with just a bit of steel guitar strumming and picking. For the first two verses, "36 Inches High" seems to be saying something pretty deep, a terse sketch of an epic life. Then in the last verse we see it's all just a ruse designed to give flight to the jokey kicker ("Once I was a ruler…/I never got over 36 inches high"). It's one of those crazy jumps in logic that songs sometimes take, which leave you going "What the hell just happened?" even after you've heard them a million times.

Nick Lowe's cover of "36 Inches High" is the version most people have heard. The song almost brings Lowe's *Jesus of Cool* (renamed *Pure Pop for Now People* in the U.S. by sensitive industry types) to a shuddering halt; it's the lone unsettlement on the man's otherwise jukebox-friendly early-career peak. While the turgid rhythm pattern and sci-fi-sounding touches of Lowe's inventive cover add flesh to the bare-bones original, my hat is off to Jim Ford for concocting this resolutely singular musical conundrum.

---

1 The number 37 has become a leitmotif in the films of director Kevin Smith. Upon learning of his girlfriend's past, Dante exclaims with horror, *"I'm 37?"* As you will soon see, #37 gets very little respect in pop culture.

A SONG TITLE	AN ALBUM	A BAND	A LYRIC
"36" Chain" – Run the Jewels (2014)	*Enter the Wu-Tang Clan (36 Chambers)* – Wu-Tang Clan (1993)	36 Crazyfists – Alaska, metal (1990s–2010s)	"When she was 36 she met him at the door" – Mary Chapin Carpenter, "He Thinks He'll Keep Her" (1992)

POSTSCRIPT: The SO36, a Berlin music club that Bowie and Iggy frequented in the late '70s, inspired "S.O.36," a song from the 1980 debut LP by Killing Joke. "S.O.36" is an apt enough representation of the band's early essence, described by its original drummer as "the sound of the earth vomiting."

37

"I, now thirty-seven years old in perfect health begin./ Hoping to cease not till death." So wrote Walt Whitman in "Song of Myself," which was surely the last time 37 was presented in a positive light in American culture. I can't confirm this regarding the century that followed Whitman's poem in 1855—sheer speculation there—but I can say that since 1967, when Lou Reed searched his mind for a number that signified utter ignominy and chose 37 ("You're written in her book/You're number 37 have a look"), it's all been downhill for the fifth lucky prime. There's "The Ballad of Lucy Jordan," for example, wherein 37 is the age at which the song's heroine "realized she'd never ride through Paris in a sports car with the warm wind in her hair." In 1975, one year after Dr. Hook & the Medicine Show had a hit with this Shel Silverstein's story-song (and five years before Marianne Faithfull's darker take) the number's fate was sealed when the 37th U.S president, Richard Nixon, began his resignation speech by noting that it was the 37th time he'd addressed the nation from that very office.

Granted, the number was treated with some measure of awe in *The Mothman Prophesies*, the last frames of which reveal the meaning of the phrase "Wake up, Number 37." But 37's usually just a goof. *The Onion* knows this well, as in, "37 Record-Store Clerks Feared Dead in Yo La Tengo Concert Disaster"[1] and "Son, You'll Thank Me For Pushing You This Hard When You're 37 and Miserable." Same deal in *Monty Python's Holy Grail*. King Arthur addresses a bedraggled figure as "Old Woman"; the person corrects him ("Man!") and then throws in, "I'm not old. I'm 37." With a cooler number— say, 35—just not as funny.

In the ramshackle "37 Pushups," Bill Callahan, operating as Smog, taps 37's dreary, end-of-the line implications and evokes a singular loneliness ("I feel like Travis Bickle/listening to "Highway to Hell"/It's a shitty little tape I taped off the radio"). You can taste the desperation in the decreasing number of pushups in this "winter-rates seaside motel." You can hear the violin wind scraping against the feeble walls, and when he says he's going up to go down again, you believe him. As for the strange cocktail that is "Black 37" by Mr. Nogatco (aka Kool Keith), it veers from cheesy sci-fi-movie dialogue to heavy-metal crunch chords and a lusty, impressionistic rap: "Her bra's made o' mink/her panties fur is a bear/My eye contact is everything I touch/I wanna lick her hair."

THE VERDICT

In 1961, social psychologist Stanley Milgram attempted to fathom the repeated assertion of the Nazis on trial at Nuremberg a few years earlier that they were just following orders. Milgram devised a series of experiments to determine just what the average person was capable of doing when reassured by a man in a white lab coat that the act was perfectly acceptable and appropriate.

---

[1] Subhead: "It's just a twisted mass of black-frame glasses and ironic Girl Scouts T-shirts in there."

The 37 in Peter Gabriel's "We Do What We're Told (Milgram's 37)" refers to the number of volunteers who were willing to administer the maximum, fatal voltage to "learners" in one of Milgram's experiments. Gabriel has always been drawn to dark subjects in his solo work, including songs from the point of view of an assassin, a housebreaker, an amnesiac, and a mental patient, and elegies for Stephen Biko and the poet Anne Sexton. With its ominous drumbeat and slowly uncoiling atmosphere, "Milgram's 37" is prime Gabriel territory. Lyrically, though, it's atypically concise from a man who once filled up album-length suites with tales of giant hogweeds. "We do what we're told" is repeated six times, followed by "One doubt/one voice/one war/one truth/one dream." And that's it. I can't help thinking that if "We Do What We're Told" had some of the detail of "Biko," the result would have been even more stunning.

A SONG TITLE	AN ALBUM	A BAND	A LYRIC
"37" – Devo (c. 1977)	*California 37* – Train (2012)	ST 37 – Austin, Texas, indie rock (1990s–2000s)	"He's got a 37-year-old guitar" – Liz Phair, "Jealousy" (1994)

38

> "That .38, you go out and hammer nails with it all day, come back and it'll cut death center on target every time."
> —Easy Andy [the gun dealer played by Steven Prince[1]] in *Taxi Driver*

GUNS

Songwriters brandish the number 38 almost exclusively as a firearm. The great Robert Johnson would not have agreed with *Taxi Driver*'s Easy Andy about the .38 ("it's most too light," in "32-20 Blues"), but his fellow bluesman Julius Daniels liked his; in "99 Year Blues" Daniels specifies "a .38 pistol on a four-five frame." Sleepy John Estes was no less emphatic on "Bring Me My .38 Pistol." The home intruder in Skynyrd's "Saturday Night Special" shoots a man "full of .38 holes," and the tradition continues all the way to hip-hop with the Notorious B.I.G. ("a sawed-off pump/chrome .38 pistol" on "Notorious Thugs") and Young Jeezy's yeyo nightmare ".38." Dr. Dre, Jadakis, and the Rev. Horton Heat also feature .38s. The title of "Small Change (Got Rained on With His Own .38)" serves as the weary refrain of Tom Waits's half-spoken, sax-addled bop song-poem. Only the hangdog protagonist in Tom Petty's "Nightwatchman" mentions having a gun permit.

BOSOMS

SONG	DIMENSIONS	KEY LYRIC
"Switchboard Susan" – Nick Lowe	"38-27-38"	"She brings a smile/to my dial"
"Bonita Applebum" – A Tribe Called Quest	"38-24-37"	"Uh uh uh!"
"Country Girl" – Johnny Otis	"Thirty-eight, twenny-fo', thirty-eight"	"Deep in the waist is her walk/ and wide in the hips as a hawk"
"Bad Boy Boogie" – Motley Crue	"38-28-38"	"I tattooed her and now she's mine"

---

1 Martin Scorsese was so taken by his gun salesman that he made a short film about him in 1978, called *American Boy: A Profile of Steven Prince*. The subject spins epic yarns about being the 21-year-old smack-addicted road manager for Neil Diamond, avoiding military service by freaking out the Draft Board, and resuscitating a junkie with an adrenaline shot, an event that inspired the crucial scene in *Pulp Fiction*. Neil Young's "Time Fades Away" plays clangorously over the credits. Watch it on YouTube. On a related note, one has to think that in Season 3 of *The Sopranos*, David Chase pays sly homage to Easy Andy's monologue when Ralph Cifaretto (Joe Pantoliano) tells his doomed stepson, Jackie Aprile, "That .38, they'll never jam on you. Plus, it's small." In *The Wire*, boxer Dennis "Cutty" Wise concurs: "Thirty-eight don't jam."

CANADIANS

Is it mere coincidence that two of the extremely limited number of #38 songs in existence are from Canadian outfits? I mean, what are the odds? (The Odds, actually, were a '90s alt-rock band from Vancouver—weird, eh?) The modern-rock twanger "38 Years Old," by Canadian Rock Hall of Fame members the Tragically Hip, concerns a dude who's reached that lofty age and never kissed a girl, while "38th Adventure" is an ingeniously well-constructed slab of glam-tinged indie rock by Hot Springs, late of Montreal, Quebec.

THE 38 CLUB:
Twenty-seven has been conspicuously fatal to rock stars, but a 2011 study concluded that 38 is just as bad. Perhaps the dead-at-38 roll call—which includes Johnny Thunders, Harry Chapin, and Brent Mydland of the Dead—simply lacks the A-list star power of the infamous original club.

THE SOUTHERN THING

It's hard to believe that there was any kind of psychedelic garage rock movement in 1960s Alabama. After all, how many parents in this bastion of segregation would let their kids make that kind of a racket, let alone do so while extolling self-expression and consciousness expansion through chemical means? Nevertheless, the revivalist label Gear Fab's *Psychedelic States: Alabama in the 1960s* provides clear evidence that there was indeed a vibrant scene, albeit composed of bands that never broke beyond brief regional acclaim. One was even led by the son of Gov. George Wallace, the human embodiment of racial prejudice at that time. Another was the Chimes, a quartet from Dothan, Alabama. Their regional hit, "38," possibly named for Dothan's county code, is a primo example of the goofy pleasures of the genre. With an unidentified singer declaiming in a cadence that splits the difference between Dylan and a Southern Baptist preacher, some breathless mouth harp, and splats of vibrating fuzz mimicking the Electric Prunes, it's a loony little shaker that should by all means be heard. And apparently it needs to be owned; the original 45-rpm single of "38" will set you back several hundred bucks.

THE VERDICT

A song like "38," by the Al Jourgensen side project Revolting Cocks,[2] reminds us that musical commemorations of horrific events need not be literal-minded treacle like "Where Were You (When the World Stopped Turning)." Written about what has come to be known as the Heysel Stadium disaster, which occurred in Brussels in 1985 during the final match of the European

---

2   Revolting Cocks ranks among the most profane band names in history. Nevertheless, the real debt owed by any band with an anatomically intimate moniker is to the Butthole Surfers. As the story goes, before settling on that infamous location, the Surfers, formed in San Antonio, Texas, in 1980, would amuse themselves by choosing a new, outrageous name for every gig they played. One night they would be Ashtray Baby Heads. Next night: Nine Foot Worm Makes Own Food. A week later it was Abe Lincoln's Bush, or the Inalienable Right to Eat Fred Astaire's Asshole. Then, one fateful night in Austin in 1982, at the last minute they introduced themselves as the Butthole Surfers, after a song Gibby Haynes had just written, and the namescape of rock was forever changed.
    So far as I can tell, no band of any stature had ever owned a name quite so offensive. Sure, Steely Dan was Burroughs-speak for dildo, and Buzzcocks, though nonsexual in origin, was still pretty in-your-face. But with a name like Butthole Surfers there was no obscure code, no ambiguity. Subconsciously or not, this name must have emboldened other bands. A year later, in the next state over from Texas, Wayne Coyne & Co. settled on the tamer yet somehow related name of Flaming Lips. Mr. Jourgensen opted for Revolting Cocks after having the epithet thrown at him by an exasperated bartender. Meanwhile, in Stockholm, garage rockers the Stomachmouths named themselves after the Swedish approximation of "intestinal valve," a term they'd found in a translation of John Kennedy Toole's *A Confederacy of Dunces*. Looking back on it, the mid-'80s was the heyday of the grotty-name trend. Having a *New York Times*-certified unprintable name takes some doing these days, although Pissed Jeans and Fucked Up are up to the task. But Butthole Surfers and their intestinal offspring were there first.

Cup, "38"[3] is no mere lament for the fans killed in that horrific melee. In fact, the message of the song, which leads off RevCo's major-label release, *Big Sexy Land*, is that this sort of thing happens all the time. Get used to it. *Dance to it.*

Now, I would never wish for all songs about disasters, floods, and levee breaks to have an industrial rock sound, but in this case the harsh, unforgiving beats match the hellish crush scenario that took place, as does the repeated "38…38/There were 38," taunting like a jackhammer. While "38" ends with the sound of a TV newsreader intoning, "I can tell you the official number of the dead is now at 38," in an odd bit of synchronicity, the Blue Aeroplanes' ominous "Police (38 Divinity)," released the same year as the Heysel tragedy, also makes use of musique concrète in commemorating a violent event. Based on a field recording of a student demonstration, it begins with the words of a tense narrator: *"Policemen with motorcycles are riding up on the sidewalks trying to run people over to clear 'em out. One cop picked up a demonstrator and tossed 'em down the steps of the subway … There goes the gas."*

A SONG TITLE	AN ALBUM	A BAND	A LYRIC
"38 1/2" – Irakere/Chucho Valdés (1998)	*.38* – Young Jeezy (1999)	.38 Special – Jacksonville, Fla., Southern rock (1970s–2010s)	"I lost 38 pounds and my eye turned to glass" – Butthole Surfers, "The Lord Is a Monkey" (1996)

LAST WORDS:

"This album will take exactly 38:49 out of your future."—back cover, The Firesign Theater, *I Think We're All Bozos on This Bus* (1971)

---

3 In the end, 39 people died; clearly RevCo went to work on this song before the final toll could be assessed.

39

Thank Tim Rice and Andrew Lloyd Webber, the men behind *Jesus Christ Superstar*, for giving the 39 lashes suffered by Jesus their own song. And they did cook up a pretty cool fuzz guitar lick to evoke the titular strokes as they are counted out, with increasingly high-pitched glee, by the actor portraying Pilate. Not the kind of song you want to pop up on Shuffle during dinner, but ideal to get you in the mood to read *Thirty-Nine Lashes—Well Laid On: Crime and Punishment in Southside Virginia 1750–1950*, by Herman Melton, which begs to be made into a concept album by Titus Andronicus.

Depending on your perspective, the age of 39 either makes you perpetually young (comedian Jack Benny always gave his age as 39), over the hill (e.g., Tenacious D's noxious ode to a not completely young woman, "39") or somewhere else entirely: André 3000 of Outkast stated, "[A]t 39, you figure out you really don't know anything." The age of 39 makes strange bedfellows of the Cure and Jerry Lee Lewis. In the Cure's "39," Robert Smith, in his plummy moan, laments, "I used to feed the fire/but the fire is almost out." In "39 and Holding," Jerry Lee Lewis employs a similarly incendiary metaphor ("He's holding to a candle/and it's burning at both ends").

While taking a lighter touch musically, the Killer's tale of a man in chronic denial of his own mortality has enough despair to please avid Cure fans—like the Catskills, N.Y.-based indie pop duo Bishop Allen, who covered the Cure's "Love Song" for their yearlong "52 Covers" project and whose songbook includes the spooky, handclap-driven "Number 39." More downbeat is the brief "Raid on Bush Creek in '39," by the unsung Goose Creek Symphony. This stark elegy for six men murdered "to please the man in the big leather chair" is from the 1970 debut by this Kentucky outfit, which has earned comparisons to the Band.

THE 39 STEPS:
In *Hannah and Her Sisters*, Canadian punkers the 39 Steps[1] virtually co-star in a scene filmed at CBGBs: the disastrous first date between Holly (Dianne Wiest) and Mickey (Woody Allen). It's the only moment in Woody's filmography where rock 'n' roll rears its unkempt head, and the unflattering light in which it and its fans are depicted says much about Woody's magical realm, where neither rock nor black people exist. Allen's antipathy goes deep. In a 2013 essay, he wrote, "I have always had an animal fear of death, a fate I rank second only to having to sit through a rock concert."

---

[1] The Thirty-Nine Steps is an organization of spies, collecting information on behalf of the foreign office of—...

David Klein

THE VERDICT

As a youth I read an interview with Freddie Mercury of Queen in *Circus* magazine, in which the singer proclaimed that the band had chosen the name Queen "for its regal connotations," and I bought it. So you might say I have always been taken in by Queen. Perhaps only Led Zeppelin rivals Queen in the sheer scope of aural territory covered, from stupefyingly heavy to fancy-pants and feather-light. By the end of Queen's reign, the inherent bombast and theatricality of the enterprise took over and the music suffered, but the first half dozen records are crammed with endlessly inventive, utterly distinctive music.

"39" still feels like the myrrh-kissed zephyr it was in the era of Boston and Kiss, cutting through the muddle with the rich warmth of acoustic strings, Land of Oz harmonies, and a narrative that soars high above pop-chart clichés. It could have been written in the 16th century. Certainly the word "grandchildren" has never been more eloquently uttered in the context of a pop song. It's also notable for being a hit for Queen sung not by Freddie Mercury but by guitarist Brian May, who composed it. May says he was tapping into a sci-fi motif—the man who comes back from space travel having aged a year, to find a hundred years have passed—and relating it to his own sense of estrangement upon returning to his native London as rock royalty. Delivering only his second lead vocal with the band, May seals the deal with his quietly assured delivery. Finally, '39" shakes out to be the 39th song of the band's recording career, a numerical naming trick that was the first of its kind.

A SONG TITLE	AN ALBUM	A BAND	A LYRIC
"The 39 Steps" – Bryan Ferry (1994)	*39/Smooth* – Green Day (1990)	39 Clocks – Germany, psychobeat (1970–1980s)	"Or can I hear the echo from the days of '39" – The Clash, "Spanish Bombs" (1980)

Ladies and gentlemen, we have struck gold. A numerical downpour of this sort has not been seen since the opening credits of *The Matrix*. Maybe we shouldn't be surprised: Big, round numbers are bound to have inspired a lot of songs, but you don't have to be on a mad numerological quest to know that 40 is a hotter decade number than 70, or even 30. So many meanings and primal events are associated with it. You can drink a 40, catch 40 winks, spend 40 days and 40 nights pining for your baby, listening to Top 40 radio while screaming along to the "Ali Baba and the 40 Thieves" section of the Beastie Boys' "Rhymin' & Stealin.'"

Musically speaking, though, 40 is usually a matter of miles, days, nights, years, or ounces. Let's start with ounces. Bill Haley's "40 Cups of Coffee" notwithstanding, 40 in song usually means malt liquor, which went global in the mid-'80s when Billy Dee Williams did his dubious St. Ides ads. Soon enough, Snoop Dog and Ice Cube were singing the praises of 40s and introducing the term to frat boys nationwide. Of course, "liquid crack" is not really a joke. Public Enemy's Chuck D., who sued St. Ides for sampling his voice in a radio ad, wrote "One Million Bottle Bags" to express his contempt for the business of targeting inner-city youth with cheap, high-alcohol brew. Nevertheless, 40s have been held up for years as objects of rap and rebellion—starting in the early '90s with Black Sheep ("Pass the 40"), Sublime ("40 Oz. to Freedom"), Bone Thugs-N-Harmony ("Bless Da 40 Oz."), and countless others. "That's why I don't fuck with the big four-oh," quoth Cypress Hill's B Real on "Insane in the Brain." But 40's association with booze goes back much further. Early on in *Huckleberry Finn*, the boy narrator tells of how his Pap "traded his new coat for a jug of forty-rod, and clumb back again and had a good old time."[1]

Forty has an Old Testament feel to it, as in the deluge that inspired Muddy Waters' stirring version of Bernard Roth's "Forty Days and Forty Nights" (and a lead-footed cover by Steppenwolf). Michael Jackson slips a 40 days and 40 nights reference into "Billie Jean," and riffs on the construct are plentiful, à la the Donnas' "40 Boys and 40 Nights." Twang king Duane Eddy's "Forty Miles of Bad Road" was covered in true surf style by the Lively Ones, as "Forty Miles of Bad Surf." Crazy, man.

Though I see several fine choices here, this contest shakes

4x40	
Mercury Rev's "Opus 40" is a rich, trippy reverie that wafts over a just-right groove by Levon Helm.	Wire's "40 Versions" is seductive, discordant, and just plain cool, with a guitar figure that loops through the song like a replicating virus.
U2's "40," based on Psalm 40 of the Bible, has been a regularly scheduled sing-along in the band's live set for decades.	Franz Ferdinand's "40 FT," is a dark tale of mountaineering gone wrong, featuring the Scottish band's trademark spring-loaded rhythm section and a rustic melody that would sound at home in the Balkans or a *Fiddler on the Roof* revival.

---

1 Forty-rod is a regional term for super-strong, cheap whiskey that was alleged to "kill at 40 rods" (a rod is an ancient measure of length equal to 5 1/2 yards).

out to be a confrontation between two classic country story-songs. "When It's Springtime in Alaska (It's Forty Below)"[2] is Johnny Horton's tale of a Klondike prospector who gets more than he bargains for when he dances with Big Ed's redheaded gal on a Kodiak rug. In fact, he sings to us from beyond the grave. Johnny and June Carter Cash remade the song in 1960, just a year after Horton died in a car accident at 35, and Cash dedicated the recording to his fallen friend. The version by Horton is charming if a bit quaint; the Cash rendition has a touch of timelessness. That voice, obdurate as an old door, next to June's demure delivery, especially on "see-loon," is heaven-blessed.

Johnny Cash was also a friend to Tom T. Hall. When he introduced Hall to American audiences on his variety show in 1968, he called the Kentucky native "a very important writer"[3] before Tom T. performed his latest single, and first Top 10 country hit, "Ballad of Forty Dollars." If "Springtime in Alaska" is a tall tale sketched in a few bold strokes, "Ballad of Forty Dollars" is a short story told with economy and quiet empathy. Hall's grave-digging narrator is very much alive and experiencing mixed emotions about the recently departed, his inscrutable widow, and the mundane stuff of life, like having to mow the grass. These and other details mesh so well with the song's sprightly country rhythm, urged on by Jerry Kennedy's dobro fills, you could almost miss the masterly way Hall conjures a recognizable human consciousness as well as a sense of place within this three-minute yarn. And that kicker of a last line—a junior O Henry ending—manages a neat trick, delivering an exquisite aha moment while still leaving things, financially at any rate, unresolved.

THE VERDICT

No story song has gained the top spot.[4] Now we have two, and they share a good deal in addition to being country songs by masterful singers of plainspoken power. Each deals with death—in "Alaska" it's a violent demise, while Hall depicts a more traditional way of departing, when a man simply lies down and doesn't get up again. They also offer contrasting versions of bad luck—there's the kind that leaves you in a financial hole and the kind that lands you in a much deeper hole. Somehow, the part of me that has to go to work and mow the grass compels me to give the nod to Tom T. Hall, teller of tales that are bigger than they seem. His words, like 40 bucks in 1968, go a long way.

A SONG TITLE	AN ALBUM	A BAND	A LYRIC
"40 Shades of Green" – Johnny Cash (2000)	*Forty* – Lee Hazlewood (1969)	UB-40 – Birmingham, England, pop-reggae (1970s–2010s)	"Well I'm 40 now/ and I'm a man" – Tom T. Hall, "I'm Forty Now" (1976)

---

2 In case you were wondering whether "Springtime in Alaska" is meteorologically sound, the weather nerds at Wikipedia assure us that "while it may be uncommon to have -40°F (-40°C) weather in the springtime, it's not impossible."
3 Hall is no doubt important, but never self-important. On the Cash show, he downgraded the title quantity to $39.95.
4 The tradition of the narrative song is beyond old, predating medieval balladry and reaching back to the time of *Beowulf*, if not the Stone Age. The songs of Leiber & Stoller—more recent exponents of early rock—were rife with self-contained tales of teenage hassles. In the late '50s the flavors varied, from the teenage pop fluff of Claudine Clark's "Party Lights" to pure corn like "The Ballad of Davy Crockett." The first No. 1 of the 1960s was "Running Bear," the tale of Native American lovers who, separated by tribal enmity and geography, go down together amid politically incorrect chants. The story-song trend continued to crash the '60s pop charts, from teen tragedy singles like "Last Kiss" (1961) and "Leader of the Pack" (no. 2/1964) through "Ode to Billy Joe" and "Ballad of the Green Berets (no. 1/1967)." Tom T. Hall's skewering of small-town hypocrisy, "Harper Valley P.T.A." (no. 2/1968), was a crossover smash in 1968. The anything-goes '70s had their share, e.g., "The Night Chicago Died" and "The Night the Lights Went Out in Georgia," but in subsequent decades, while still thriving in any number of genres, true story-songs have been scarce in the upper reaches of the pop charts. Eminem's yanked-from-the-lurid-headlines hit "Stan" comes to mind, as does "Love Story," Taylor Swift's co-opting of *Romeo and Juliet*. But let's face it, these days, lyrical coherence and subtlety has taken a backseat to the science of pop hooks, bass drops, and blow-your-mind production.

Enunciations of "forty-one" are rare things in music; the pinnacle of these might be the moment Tom Petty's heroine in "American Girl" hears the cars roll by "out on 441 like waves crashing on the beach." It was rumored that the song memorialized a woman who committed suicide at the University of Florida, but Petty has emphatically refuted the notion that he was referring to anything other than U.S. Route 441, which begins in Miami, passes through his hometown of Gainesville, Fla., and winds north to Tennessee.

I've never been much of a map-reader, but I've always dug the way Tom spits out those numbers.[1] Nearly 30 years later, T.P. returned to the scene in "Route 41," named for the major north-south route of which Route 441 is a stub. With its stripped-down feel and a distorted vocal effect reminiscent of Zeppelin's "Hats Off to (Roy) Harper," this raw blues is one of the high points of the less-than-stellar *Mojo*.

Iron & Wine merchant Sam Beam and Calexico make reference to the same road on "Prison on Route 41," an evocative waltz-time tale of a man who avoids the fate of incarceration suffered by his family members, thanks to his love for "the righteous grand Virginia." Of several numerical titles in the Dave Matthews Band catalog, "41" has the greatest mystique. The band treats it as a set piece in concert, an opportunity to jam extensively with "special guests." (The longest performance thus far lasted over a half hour.) The subject matter of "41"—a dispute over song ownership—is strangely apropos: it was originally titled "41 Police," due to its similar feel to "Bring on the Night" by the Police, which, as recounted earlier, served as the uncredited DNA of "Edge of Seventeen."

In "American Skin (41 Shots)" Bruce Springsteen lamented the death of Amadou Diallo, the 23-year-old native of Guinea who was met with a barrage of NYPD bullets in 1999 for no good reason. Springsteen earned the wrath of the law-and-order types in his fan base by writing this stark and affecting elegy, which went viral on the Internet without an official release. "41 Shots" is something to be played sparingly, in the same way that even the most diehard Spielberg fans reach for *Raiders of the Lost Ark* more often than *Schindler's List*—the soundtrack of which contains another dark #41 song: "Jewish Town (Krakow Ghetto – Winter '41)."

THE VERDICT

Some songs you love, and you carry them around in your head like treasures. But some have a different kind of power—they hold you in their thrall. You can carry them around in your head, but still you're almost a little afraid of how good they are; you feel the

---

[1] This map reference is what we numerologists refer to as foreshadowing.

way *Sopranos* heavy Bobby Bacala did when he told Uncle Junior, "I'm in awe-r of you." For me, "Map Ref. 41°N 93°W" by Wire is such a song. I'm in awe-r of it.

Despite a melody that entices, a soaring chorus, and a futuristic-sounding sonic palette, it's still a bit of a glorious blur to me, like a rainbow in a puddle that disappears when I try to grab it. The specificity of the title and the clearly enunciated attack of the main guitar lines are at odds with the song's overarching elusiveness. The coordinates in the title, after all, make specific reference to the terrestrial equivalent of nothing at all: a field in Iowa. That same elusiveness and the overall smeared quality of this 1979 song became hallmarks of My Bloody Valentine a good 10 years later. And as far as I know, no one else but MBV has had the guts to cover it.

A SONG TITLE	AN ALBUM	A BAND	A LYRIC
"Forty-One" – Califone (2005)	*41* – Swell (1993)	Sum 41 – Ontario, Canada, punk-pop (1990s–2010s)	"I was born in the backseat of a Greyhound Bus/ rolling down Highway 41" – The Allman Brothers Band, "Ramblin' Man" (1973)

42

In *The Hitchhiker's Guide to the Galaxy*, the number 42 signifies the meaning of life itself. The members of Level 42, London-based purveyors of vaguely danceable synth-pop in the '80s, were so taken with Douglas Adams's sci-fi behemoth that they invoked it when naming their eventually quite successful band. In the world of song, 42 stands for something only slightly less fraught with possibilities than life itself: 42nd Street.

Here's the strange thing; in a certain sense, there's no mystery about the definitive #42 song: it's "42nd Street." You know the one: *"Come and meet those dancing feet/On the avenue I'm taking you to/Forty-second Street."* But according to The Rules laid out at the start of this trek, show tunes are out of bounds because they exist outside of the rock realm. They do incorporate some of rock's flavors—it's easy to connect the dots between Tin Pan Alley and Sir Paul's more winsome Beatles songs. And Bob Dylan wrote in *Chronicles Vol. 1* that something vital clicked for him as a youth while watching a performance of Kurt Weill's *Threepenny Opera*. To be sure, some of my favorite performers—the Kinks, Bowie, Kate Bush—have more than a hint of West End to them.

But a song like "42nd Street," whether the Depression-era ditty by the Boswell Sisters or the belted-out Broadway showstopper, just doesn't make sense at this particular party. My inability to turn up a single rocked-up or bluesy version of "42nd Street" seems to bear this out, and the fact that deep digging has only yielded several bad "42nd Street"-titled songs (by Flaming Lips, Golden Earring, and Piper) reinforces the idea. As I contemplate the original "42nd Street" sitting cheek by jowl with "Map Ref. 41°N 93°W," my inclination is to take a strict-constructionist reading of the Rules' penultimate proscription: "Rock and pop, R&B, blues, soul, funk, hip-hop, country, and folk are the key genres under consideration."

This also leaves out "Psalm 42," a 12-minute symphony of sorts by the Trees Community, a monastically minded troupe whose first release, *Christ Tree* in 1975, was too baffling for widespread recognition. Yet the absolute singularity and authenticity of the project has led to a wave of belated positive reappraisal. The sonic landscape that opens "Psalm 42" could almost pass for Eno's *Another Green World*, but it doesn't last long; soon vocal interplay enters, suggesting plainsong, then Eastern bells and oboes, and voices chanting songs of praise—and that's only in the first half.

Many Coldplay-heads (Head Colds?) conjectured that Coldplay's "42" is *Hitchhiker's Guide*-related, but lead singer Chris Martin dispelled that notion with a highly un-cosmic explanation: "It's called '42' because it's my favorite number." Yet speculation continues to run rampant, at least among the Songfacts.com faithful.

"I think that there is a Christianity reference somewhere in that," says Benjamin of Birmingham, Ala., of the line "There must be something more…" Meanwhile, Jack from Detroit[1] has other ideas: "I read the band were hypnotized before they recorded *Viva La Vida*. I have a hard time not believing that the band channeled the lyrics. 42, could it be related to Orion's Nebula, M42?"

---

[1] A fictional Jack—Jack Torrance—is the psychopathic heart of Stephen King's *The Shining*, the film version of which is a hotbed of 42-ism: Jack takes 42 Louisville slugger swings at the hapless Wendy, *Summer of '42* is playing on the TV, Danny wears a #42 T-shirt, etc. As a result, certain film obsessives have concluded that Kubrick's adaptation, with its numerous subliminal references to '42, is a disguised Holocaust parable.

David Klein

THE VERDICT

East River Pipe is the musical alias of Fred Cornog, a reclusive yet prolific songwriter whose weary voice hints at the hard row he's hoed. After a brief flirtation with major labels in the early '90s, followed by years of homelessness and drug addiction, Cornog has found stability while continuing to write songs marked by understated beauty and a wry lyrical touch. It took me a listen or two to fall for the simple charms of "Down 42nd Street to the Light," but I now see its strengths clearly: the mix of hope and resolve, the ramshackle but just-right musical accompaniment, and the hypnotic singsong of the secondary vocals, like a child's voice issuing from a car's backseat at dusk.

Before I discovered the East River Pipe song, I wondered if I might have to just bite the bullet and choose Django Reinhardt's "Swing 42," a spry instrumental by the three-fingered, trailblazing guitarist. After all, Django is my dad's favorite guitar player, and his gut-string strumming is as familiar to me as birdsong. But the thing that proved I had found a #42 song I could really live with is that Cornog mentions my hometown in the second verse: "We could fly from here to there and back/Tenafly or maybe Hackensack."

Now, let me assure you, references to good old Tenafly—also the hometown of Ed Harris, Leslie Gore, and Bob Guccione Jr.—are few and far between in pop culture, and they are something I savor.[2] It was all the sign I needed. Yet there was an additional sign, which was just plain odd: a 1995 release by East River Pipe was called *Poor Fricky*.

My first pet was a cat named Fricky.

You can ask my dad.

A SONG TITLE	AN ALBUM	A BAND	A LYRIC
"42 in Chicago" – Merle Kilgore (1961)	*Summer of '42* – Tony Bennett (1972)	Visitor 42 – San Francisco, punk (1990s–2000s)	"Sign page 42/ we'll do the rest for you" – Warren Zevon, "Even a Dog Can Shake Hands" (1987)

---

2 Like Roz Chast's 2012 *New Yorker* cover depicting a fantasy Second Avenue subway line on which Tenafly has its own stop, between Mt. Fuji and Papua New Guinea.

43

I did say 37 was the number everyone loved to goof on, but the same goes for 43, which has been manna to comedians for decades at least. In Monty Python's fictitious game show *Stake Your Claim*, a contestant named Voles alleges he has authored all of Shakespeare's plays. When the host asks his age, Voles replies, in a flat, reasonable voice, "*forty-three*"—and it's somehow funnier and more ridiculous than if he'd said 33 or 48. The singular inelegance of 43 is the comic DNA of 43-Man Squamish, a sport dreamed up by *Mad* magazine's George Woodbridge and Tom Koch in 1965 that commences when the words "My uncle is sick but the highway is green" are uttered in Spanish. More recently, Louie C.K., playing himself in *Louie*, is confronted by the gal he can't get, who explains his plight with the words, "Effin *forty-three*, Louie." And then there's Patton Oswalt, jock-rocking the tragedy that is KFC: "Eatin' my lunch from a single bowl/ Happy birthday, I'm forty-threeeee..."

So is the number cursed? Put it this way: The past few years have seen the release of films called *42* and *Movie 43*. The former, an inspirational story of Jackie Robinson, earned some critical plaudits as well as its money back; the latter was dubbed "the *Citizen Kane* of awful" by *Chicago Sun-Times* reviewer Richard Roeper. I rest my case.

One of the 43 most difficult things to explain to people who weren't around in the 1970s is how a band led by a shaggy-headed, codpiece-wearing man in tights, given to playing the flute while balanced on one leg, could regularly play to a sold-out Madison Square Garden before throngs of adoring teenagers. How indeed could a combo named after the 18th-century British agriculturist who invented the seed drill grow hugely popular, in the States no less, playing complicated songs that drew heavily on Irish jigs and 'round-the-maypole reels? It sounds ludicrous now, doesn't it, like Spinal Tap's dancing elves?

The biggest mystery is what it all meant. Most bands gave the youth something to go on—"Love is all you need"; "You don't need a weatherman to know which way the wind blows"; "Well, whatever, nevermind"—but what was the proper response to Jethro Tull? Wear a filthy trench coat? Take up archery?

It's tempting to say a certain type of nonsense went further when our world had yet to be digitized, but what really made the band relevant to the kids is quite simple: the music's heavy rock foundation. The rock coexisted with the Olde English folke traditions without harshing anyone's mellow. Even as Ian Anderson's look went from the mad beggar of *Aqualung* to the Robin Hood duds of *Songs From the Wood*, there were still plenty of loud guitars. And in the early '70s, young people in many quarters liked music that had the appearance of depth. They had no problem with artifice and couldn't have cared less about "authenticity" or whether you could dance to it. And Tull always seemed to be saying something deep. "Aqualung," a mini-symphony about a pervy old coot, would not have struck the same chord if said coot weren't called Aqualung, and instead went by, say, Old Scratchy. With its brawny guitar riff, tailor-made to be aped by teenage boys "nare-nare-nare"-style, and a strategic reference to snot, it's a song with eternal appeal to the young at heart.

As for Tull's #43 song, "Hymn 43" shows off guitarist Martin Barre's knack for a memorable hook as well as Anderson's distinctive delivery. The song was part of a growing trend in early-'70s rock of singing about Jesus Christ; Lloyd Webber's *Superstar* premiered on Broadway, and the harvest was plentiful with songs like "One Toke Over the Line" (sweet Jesus!) and the Doobie Bros' cover of "Jesus Is Just Alright."

Country Joe and the Fish became slightly notorious for leading the muddy masses at Woodstock in a chant of "F-U-C-K." Those onstage antics, and the band's jokey name, may not suggest they had the depth to record the ambitious seven-minute instrumental "Section 43." But this dark, organ-drenched psychedelic suite, written a few years before bands like the Grateful Dead made lysergic jams commonplace on vinyl, has been hailed as one of the best and truest examples of acid rock.

Being unfamiliar with the mathcore genre, which combines the relentless attack of hardcore with math rock's byzantine precision, my reaction to Dillinger Escape Plan's "43 Percent Burnt" was simply to run and hide. It's such a brutal, impossibly complicated cluster bomb that, next to it, "43" by neo-heavy metalists Mushroomhead is a Stone Temple Pilots ditty, and the anti-pedophile "Rule 43" by Glaswegian oi purveyors Bakers Dozen is a lost Proclaimers song written after a lager fight.

The lush, unabashedly romantic songs of the bird-obsessed U.K. outfit known as the Guillemots might come off as a bit too sweet if they weren't shot through with intriguing instrumentation, plus hints of conflict underneath the often-soaring surfaces. "Made Up Love Song #43" showcases these qualities. If you can get past the opening reference to "shining dragons," after the second chorus the whole thing bursts into bloom, with especially lovely interplay in the rhythm section and over-the-top falsetto background singing. The song is imbued with the irrepressible optimism of the truly smitten. Only a person nestled deep in the arms of *amor* finds "poetry in an empty Coke can."

THE VERDICT

Mary Lou Lord took her experiences busking in London and Seattle to great indie rock heights in the late '90s, earning praise from Elliott Smith, the resentment of Courtney Love, and releasing a small but piquant selection of EPs on the Kill Rock Stars label. Her second major-label release included "43," which she sings as a whispery confession that recalls the work of the aforementioned Mr. Smith. Through a popular social media outlet I contacted Ms. Lord about the song, and she was kind enough to respond.

"It was written by my friend, sometimes co-writer, and song genius Nick Saloman of The Bevis Frond. It's a song about a young man (or woman) who is 17 having a relationship with someone who is 43. *Summer of '42*? *The Graduate*? ...you know. What's cute about it is that she doesn't call him by name ... she just calls him '17.' It really is a beautiful song."

A SONG TITLE	AN ALBUM	A BAND	A LYRIC
"43" – Vanilla Trainwreck (1992	*Songs From Motel 43* – Knowlton Bourne (2015)	Cold Forty Three – L.A., pop-punk (2000s–2010s)	"If you'd been in the SS in '43/ You'd have been kicked out for cruelty" – The Pretenders, "I Hurt You" (1984)

44

"I'm telling you, son, you know it ain't no fun starin' straight down a .44," warned Ronnie Van Zant in Lynyrd Skynyrd's "Gimme Three Steps." Indeed, 44 has the distinction of being more than just a quantity—it's an actual object: a .44—like a .38, a 40oz, and a .45. Packing a .44 has a storied history. There's a sheriff on Woody Guthrie's trail "with a big 44" in "Bed on the Floor," and there's a ".44 smokeless" in Dylan's "Little Sadie." Not to mention the *po-lice* in New York City chased a boy right through the park (with their .44s) in the Stones' "Heartbreaker." (It's true that these figures are fractions of one, not true 44s, but the decimal point is never pronounced. Let's not get crazy here.)

The .44 holds a pivotal place in the ur-folk song "Stagger Lee," which dates to an 1895 St. Louis barroom murder and came to prominence in 1922 with Mississippi John Hurt's version. It has since been sung in hundreds of ways, but "Stagger Lee" almost always includes a .44, occasionally a .45. Surprisingly, the mighty Howlin' Wolf never covered "Stagger Lee," but apparently he never left home without packing his piece. In his take on Roosevelt Sykes's "Forty-Four," which has been covered by Little Feat and the Kills, Wolf delivers an impassioned vocal in his unmistakable jagged-edged timbre. "I wore my .44 so long," he wails, "I've made my shoulder sore." Carey and Lurrie Bell's "I'll Be Your .44" is that rare invocation of the .44 as a symbol of love and devotion.

PALATE CLEANSER #44
"The forty-four seconds preceding the greatest drumroll ever recorded" refers to the Cars' "Just What I Needed," according to Rob Harvilla's essay in *Marooned: The Next Generation of Desert Island Discs*.

THE VERDICT

The Zombies' *Odessey & Oracle* is recognized as one of the finest pop records of its, or any, decade, but in 1968 only Al Kooper seemed to appreciate its greatness. Kooper, the organ player who had made history with Dylan *twice* in 1965—at the Newport Folk Festival and on "Like a Rolling Stone"—became a sought-after session man, playing guitar and keys with the Stones and Cream, Hendrix and the Who, and many lesser mortals. In short order he started, and subsequently ditched, two successful bands of his own—the Blues Project and Blood, Sweat & Tears—and by summer of '68 he would soon add "solo artist" to his résumé. So when Kooper took a summer holiday in London, he must have felt like he was floating on air. Being an extremely in-demand session musician was all well and good, but he was ready to take the next step.

David Klein

When Kooper returned to New York City that fall, to his job as an A&R rep for Columbia Records, he brought back a stack of British records that he'd picked up during his visit. Gradually, one of them started to haunt him. *Odessey & Oracle*, Kooper would write, "stuck out like a rose in a garden of weeds." Kooper got so enthralled that he personally leaned on Clive Davis, the new head of CBS Records, which owned the LP, to release it in the States. CBS didn't think much of the record's commercial prospects, nor of the five wispy-looking Englishmen and their intricate, utterly English-sounding psychedelic pop. But Kooper, with nothing to gain from his efforts save for avoiding a crime against humanity, persuaded Davis to change his mind.

Odessey was released in mid-'68 in the U.S. with little fanfare, and no one paid much attention to it or its first single, "Care of Cell 44," which had already flopped in the U.K.[1] Nevertheless, it's a perfect pop tapestry woven of Colin Blunstone's dreamy vocals, washes of gauzy Mellotron, harpsichord plinks, and a McCartney-esque melodic bass line. The exquisite, guitar-less arrangement just soars. In addition to its remarkable beauty, the song offers a scenario that's extremely rare if not unique in the annals of rock: a love letter to an incarcerated girlfriend. In the hands of Johnny Cash or Nick Cave, this lyrical conceit would be a dirge, but the Zombies fill it with such barely suppressed joy and musical inventiveness (the falsetto-dominated middle eight is remarkable) as to render the prison part irrelevant. We know she's a good gal; it's probably just a lot of parking tickets or something. The point is she's coming home, and soon, and we can hardly wait.

A SONG TITLE	AN ALBUM	A BAND	A LYRIC
"44" – Fats Domino (1952)	*44* – The Chocolate Watchband (1984)	June of 44 – U.S., post-hardcore (1990s)	"I stuck that lovin' .44 beneath my head" – Hank Thompson, "Cocaine Blues" (1959)

---

[1] The dirge-like follow-up single, "Butcher's Tale (Western Front 1914)," fared just as poorly. By the time single no. 3, "Time of the Season," became an international hit in early 1969, the Zombies had long since broken up.

45

The seven-inch 45-rpm vinyl disc is the medium that delivered rock 'n' roll to millions of teenagers in the '50s and '60s. So, numerically speaking, 45 is royalty. A spinning 45 was a familiar object of love and devotion for two generations of American youth, and then, thanks to the opening sequence of *Happy Days*, it came to symbolize the early rock era itself. By the time "Brimful of Asha on the 45" became an inescapable refrain at dance clubs in the late 1990s, the singular connotation of 45 meant nothing to the casual pop music listener. But in 1981, the term still had meaning, if the success of "Stars on 45" by Stars on 45—a medley of mostly Beatles songs set to a boom-splat disco beat—is any indication.

Stars on 45 was not a real band, just a bunch of studio musicians taking cues from a guy named Jaap Eggermont, who had devoted much time and energy to a project that had proven a nightmare to assemble.[1] Primarily voiced by fake Paul McCartney (Okkie Huysdens), fake George Harrison (Hans Vermeuien), and fake John Lennon (Bas Muys), "Stars on 45" ascended to the top of the U.S. charts a mere six months after Lennon's murder. (It would take 25 years and the strenuous intervention of Cirque du Soleil to render the Beatles this unpalatable again.) Many found the singing soulless, the beat mind-numbing, and the medley form wanting, yet "Stars on 45" spawned a short-lived revolution. It wasn't just novelty purveyors like Weird Al Yankovic ("Polkas on 45") and the British Weird Al, Ivor Biggun ("Bras on 45") who bought a ticket to ride on the medley train; even legitimately cool bands like Squeeze ("Squabs on Forty Fab"), the Damned ("Damned on 45"), and Orange Juice ("Blokes on 45") got in on the act, too. A few decades later, LCD Soundsystem hoisted the medley to rarely reached heights on "45:33," which stitches together some of James Murphy's most terrific grooves into a jog-worthy whole. Nevertheless, "45:33" is an album masquerading as a song, and the fact that it's priced as such on iTunes proves my point.

Songwriters tend not to reach for a .45 pistol as often as a .44, but the gun that made the Colt Company famous (aka the "Peacemaker" as noted by Steve Earle in "Devil's Right Hand") has had its moments in songs by L7 ("Ms. 45"), Three Six Mafia ("Paul With Da 45"), and Bad Company ("Smokin' 45"). "Colt 45" by Metal Urbain

---

[1] Stars on 45 was not the only 45-related venture in Jaap Eggermont's career: as the drummer for Golden Earring, he played on the war-themed "Another 45 Miles."

is an appealingly reverb-laden rave-up that gives French punk a good name. Shinedown had a big hit in 2003 with "45," a slice of packaged angst with a testosterone-fueled chorus that goes, "And I'm staring down the barrel of a 45/Swimming through the ashes of another life." But hang on—it's not what you think: According to singer Brent Smith, "[T]he 45 isn't an actual literal term for a gun. I used it as a metaphor for the world." Maybe so, but don't tell Bronson Arroyo. The Cincinnati Reds pitcher (and decent guitarist) almost certainly chose "45" as his entrance music because it inspires him to go out there and be aggressive early in the game, not for its metaphorical implications.

Metaphors are grown-up thoughts, and 45 is a grown-up age. Somewhere around 45, it becomes incumbent upon a person to give at least a passing thought to his or her mortality. In 1955 the Irish soprano Mary O'Hara sang the longevity-minded "45 Years." Ms. O'Hara's name may not be familiar in these parts, but her life has been made into a play, and for good reason: twice she achieved fame as a recording artist, separated by 12 years of living in a convent. If that doesn't scream biopic I don't know what does. Is Holly Hunter available?

45 is the name of Bill Drummond's collection of cranky tales about life and the music business. Drummond, whose musical sojourn began in the early '80s behind the scenes of Echo & the Bunnymen and the Teardrop Explodes, went on to score worldwide hits with the KLF and notoriously burned a million English pounds in 1994. (He now regrets it.) The book is an intermittently fascinating account that veers between fanciful discourse on interstellar ley lines and brilliant, punchy writing, like this thumbnail description of Bunnymen guitarist Will Sergeant (circa '78*)*: "Short-order chef with black moods and beautiful eyes. Favourite Stone: Brian Jones."

Julian Cope, leader of the aforementioned Teardrop Explodes, gives us a powerful statement of geophysical weirdness that aligns nicely with Drummond's ley lines in "Upwards at 45 Degrees." Opening with a two-chord progression that echoes that of Bob Dylan's "Hurricane" and evokes a similar dire feeling, Cope unspools a grandly psychedelic vision of a mother ship's arrival: "Four hundred meters across/hanging like a football field." A few verses in, Cope almost loses the plot, but he builds it all back up with slow-burning intensity that matches the song's galaxial intentions. *Jehovahkill* is the middle of an early-'90s trilogy of records that stands as Cope's creative peak. In 2000 he revealed in an interview that he regarded "Upward at 45 Degrees" as his favorite among his own songs. Scott Miller, in *Music: What Happened?*, called it "a dream of blasting out of regimented, Christian terrestrial life," though even he admitted to being unsure of whether it's a metaphor or what.

"45 GRAVE" BY 45 GRAVE
In naming a song after themselves, these '80s West Coast goth punks (led by mainstay Dinah Cancer— say it out loud) partook in a hoary tradition, one that unites Talk Talk, Motörhead, Black Sabbath, the Monochrome Set, Bad Company, Built to Spill, Dots Will Echo, Veronica Falls, Talulah Gosh, Big Country, Madness, Kaleidoscope, The Descendents, Minor Threat, Titus Andronicus, NWA, Gov't. Mule, Jilted John, Bo Diddley, Aphex Twin, Combat 84, Night Ranger, Slipknot, Meat Puppets, Damn Yankees, the Blue Aeroplanes, and Living in a Box.

THE VERDICT

Leave it to Elvis Costello, wordsmith nonpareil, to hit all the major connotations of the number in one song: 45 the year, 45 the 7-inch single, 45 the gun—and he wrote it at the age of 45. The lead track from *When I Was Cruel*, "45" marked a return to the kind of music E.C. hadn't made since he was 25. Gone, at least for the moment, was Elvis the UCLA artist-in-residence and Anne Sophie von Otter collaborator; back after a long absence was the seductive, bitter, guitar-strumming Elvis who made himself essential with an audacious vinyl troika in 1977–79. That voice is still that voice, the lyrics still sting, and the guitar crunch hasn't aged badly at all.

A SONG TITLE	AN ALBUM	A BAND	A LYRIC
"Forty Five 45s" – Shrag (2009)	*Super 45* – Stereolab (1991)	Scars on 45 – Yorkshire, England alt-rock (2010s)	"I know I was a 45 percenter then" – The National, "I Need My Girl" (2013)

"46 DAYS"–PHISH (2002)	<— CONTRAST & COMPARE —>	"PSALM 46"–TREES COMMUNITY (1975)
Stoner fable	SONG TYPE	Spiritual parable
"[T]he devil's drawing near"	TELLING LYRIC	"The Lord of Hosts with us"
"[an] interstellar quantum-tunneling rock bulldozer" (Phish.net)	SOUNDS LIKE	"the music of Claude Debussy, Bela Bartok, and Charles Ives" (thetreescommunity.com)
Religious following	DEVOTIONAL ASPECT	Religious collective
Improvisation	CONDUCIVE TO	Contemplation

THE VERDICT

I know what you're thinking: "39-21-46" by the Showmen lacks the ideal configuration for the #46 slot. It should go in the #39 slot. Well of course it should, but that's taken, and taken well, by Queen, and we need a #46 song now. There's no denying the list would scan better if "46" came first, as it does in "Forty Six & Two," Tool's mini-opus about mankind's ascendancy to a higher level of existence via two extra chromosomes. But with a song this sublime and timeless and smile-inducing, we need to be thankful for either a printing mix-up or some record-company chicanery that led to the original 45-rpm of this single being eligible for inclusion herein in the first place, giving us not just a valid #46 song but a song that should be taught to babies and old people and everyone in between.

The record is really called "39-21-40 Shape," and it's clear to the naked ear that General Norman Johnson, who wrote and sang it, sings the title that way, with nary a 46 in sight. He believes the title was deliberately changed by execs at Minit Records as a ploy to "arouse curiosity." That makes sense; it would be hard to imagine someone really mishearing "forty shape" for "forty-six," and it was a common practice among labels to change the names of songs—even the names of acts they controlled—at their whim. Johnson's own group had been called the Humdingers until Minit changed the name to the more upscale Showmen. And, on a more practical level, even to those who like 'em big, most would agree that 46-inch hips stray from the feminine ideal.[1] The hips that the song celebrates are still ample, just not 46-inch ample. Johnson croons in a voice filled with lilt and longing ("*Yoo-o-o-o-o/ with your thirty-nine twenty-one forty shape/you got me going ape-ity-ape over you*") before erupting into a joyous vamp, leading to a fadeout that leaves you wishing for more.

In the summer of 1963, that "mislabeled" single became a huge hit on the jukeboxes of Myrtle Beach, S.C., the hotbed of the Carolina "beach music" scene, where the hip white kids went to do the Shag and listen to forbidden "race" music. The Showmen were the kings of the scene. Eventually they became the Chairmen of the Board, best known for "Give Me Just a Little More Time" but makers of many more great songs courtesy of the legendary Detroit team of Holland-Dozier-Holland.

"39-21-46" falls squarely into a tradition of songs, like Sonny Boy Williamson's "Eyesight to the Blind," that attest to the healing powers of feminine pulchritude. In "39-21-46" Johnson imparts to us in his distinctive moan, with every fiber of his being, that his voluptuous heroine can make a crippled man walk and a blind man see. The interplay between the lead vocal and the doo-wop style accompaniment is an irresistible tribute to the divinity of women, one that calls to mind a verse from the Book of Talking Heads—"The world moves on a woman's hips"—or these lines from e.e. cummings' "my smallheaded pearshaped":

whereas the big and firm legs moving solemnly
like careful and furious and beautiful elephants

(mingled in whispering thickly smooth thighs
thinkingly)
remind me of Woman and

how between
her hips India is.

A SONG TITLE	AN ALBUM	A BAND	A LYRIC
"46 Satires" – Besnard Lakes (2013)	*46* – Kino (1983)	Forty-Six – Texas, pop-rock (2000s–2010s)	"Come get your kicks/on the corner of Lincoln and forty-six." – Ricky Nelson, "Waitin' in School" (1957)

---

[1] Then again, in his paean to feminine amplitude, "She Got to Wobble When She Walk," Sugarboy Crawford sings, "65 in the hips/ 55 in the waist/ a long lean gal ain't worth doodly-squat."

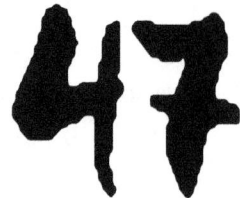

Forty-seven is as nondescript as they come, yet it has a pair of high-visibility moments in the early rock canon. In "Jailhouse Rock," Elvis calls out, "Number 47 said to number 3/you're the cutest jailbird I ever did see" in a way that 46 or 48 can only envy. Bo Diddley also spoke the 15th prime memorably ("I walk forty-seven miles of barb wire/Got a cobra snake for a necktie") in "Who Do You Love."[1]

Standalone 47s are rare birds though. Finding a #47 song—one that I could believe in—proved to be a challenge. "PO Box 9847," the Monkees' version of "Want Ads," was not eligible, although it was surpassingly stupid and catchy. Mark Kozelek of Red House Painters was certainly eligible for "Metropol 47," a heartfelt if lugubrious love song in which he sings about his desire to kiss his beloved's "sweet koala face." The rollicking "47th Street Boogie" by legendary blues pianist Memphis Slim and his hero, Roosevelt Sykes, praises New York's 47th Street as a place where you'll meet the hepcats and the fly chicks, as well as get your solid kicks.

THE VERDICT

Feeling a bit desperate, I dug around in my vinyl collection and turned up something promising, and that discovery led me to an even better one. Funny thing was, both songs were by Dwight Twilley. This struck me as curious, and when the usual online sources failed to supply an answer, I realized it was up to me to find out why Dwight Twilley had written a pair of #47 songs.

I was actually kind of nervous. You don't often speak to someone you know exclusively through your record collection. Good thing that when I called him, Twilley didn't mind explaining it to me. We got right down to the matter at hand.

"I think it's a sexy number," he declared. "You know, when you just say it, the way it rolls off the tongue. It has great syllables."

True to that spirit, "Rock and Roll '47" captures what a man sounds like when he is truly enamored of a number. Dwight stretches it out like he doesn't want to let go of it: "forty *seh-heh-heh heh-eh-vunnn...*" echoing Buddy Holly's "a-*weh-aheh-aheh*" intro to "Rave On." Given how much Twilley digs 47's mouthfeel, one might conclude that he's using it here solely for the sound, but he confirmed it had another meaning.

"That came from the musician's union in Los Angeles, which used to be called—and maybe it still is—Local Union Number 47." [It still is.] "Because, well, that was kind of the point of it. Like, this was just another rock 'n' roll song. It could have been 46. It

---

1 Operating on a more theoretical plane, Beardless Harry rode the trolley down to 47 in the Velvet Underground's "Run Run Run."

could have been 45. Could have had a name or not had a name. Coulda been a bit more up-tempo or slower. But it's just another rock 'n' roll song."

When Dwight first began making records, the "just another rock 'n' roll song" aesthetic still had legs. Rock was, after all, a familiar idiom, and, even though it had been turned into something complicated by a lot of progressive outfits, bands like Cheap Trick, the Raspberries, and Tom Petty & the Heartbreakers (Twilley's label-mates at the troubled imprint Shelter Records) were more interested in mining rock 'n' roll for its primal pleasures. They wrote catchy, Beatles-influenced songs featuring tight harmonies and sharp guitars. Most were about girls. Dwight distinguished himself through his fondness for the rockabilly "slapback echo" effect, which gave his vocals more than a touch of Sun Studios-era Elvis; "Rock and Roll '47" exemplifies this early sound.

But the stunning title track from *47 Moons*, Twilley's 2005 album, is another thing entirely. It's a song most definitely made by a grownup, with sumptuous Spectorian production (engineered by Dwight's wife, Jan), an indelible minor-key melody, a gorgeous guitar excursion courtesy of longtime Twilley guitarist Bill Pitcock IV, and a rich sense of longing and melancholy that puts one in mind of the Righteous Brothers.

"I was driving at night," Twilley told me, "and I tuned into one of those late-night radio shows—you know, where they talk about UFOs and zombies and stuff. This particular show, they had a scientist on, a real specialist, and so it wasn't so much fiction but scientific-oriented. And he just happened to matter-of-factly point out that Jupiter had 47 moons, which immediately caught my attention. And it's kinda like, doesn't seem fair. We only have one. And obviously, with the word 47, it was just a natural for me.

"I spent a considerable amount of time working on it, because I got real serious about it. And then, coincidentally, about a week later I had finished the song—or I thought I had finished the song—and I open up the newspaper here in Tulsa, through the science section, and there's a big headline that says, 'More Moons Discovered Around Jupiter.' So I had to go back and add another verse: 'They thought that there were forty-one/They'll find a thousand before they're done.'"

A SONG TITLE	AN ALBUM	A BAND	A LYRIC
"New Mission Terrace No. 47" – Moth Wranglers (1999)	*Object 47* – Wire (2009)	Black 47 – New York City, Celtic rock (1980s–2010s)	"Page 47 is unsigned/ I need it by this evening" – The Church, "Destination" (1988)

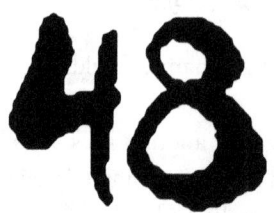

48

For most of the first half of the 20th century, "the 48" signified the continental United States, as in "Let's Get Away From It All," the pop standard in which Sinatra vowed to repeat "'I love you, sweet' in all the forty-eight." It was such a familiar and comforting notion that Howdy Doody, America's first puppet celebrity of the TV age, sported 48 patriotic freckles on his lacquered wooden visage. The great musical iconoclast Spike Jones, who added gargling, whistles, and a large dose of nuttiness to pop standards and the classical canon, made use of this association in the surprisingly unfunny "Forty-Eight Reasons Why." Laying out 48 reasons to heed the call of Uncle Sam (one state = one reason), it ends with a caffeinated recitation of state names amid bugle riffs and the sound of marching feet. The result is as heavy-handed and forced as it sounds.

"Forty-Eight Portraits," the 17-minute suite that closes Sloan's *Commonwealth*, certainly deserves consideration. Given the directive of coming up with his own *side* (as if that makes sense anymore) for the 11th studio album by this four-man Toronto-based band, drummer Andrew Scott decided to do something in the spirit of Side 2 of *Abbey Road*. You know, because it was there. The title refers to his love of painting and print-making. And while it takes a good three minutes of barking dogs and drifty keyboards to get into the song proper, "Forty-Eight Portraits" is intriguing throughout its length, shifting from tremulous psych-pop to a beguiling children's chorus section and a culminating "She's So Heavy"-esque chord cycle. But I'm going to have to invoke the "45:33" precedent, established three chapters ago, which says if a song is not available as a one-dollar download, it's more like a mini-album and not a song in the true sense. But don't let that stop you from checking out Sloan's remarkable, underappreciated discography.

The James Gang's "Funk #48" features the same kind of crunchy Joe Walsh guitar licks that make "Funk #49," which followed a year later, so recognizable. "Funk #49" is clearly the superior song—stronger melody, more interesting vocal flavor—but "48" is no slouch. The band has an intuitive grasp of the looseness-to-tightness ratio that makes a rock trio such an ideal vehicle to deliver the goods. What the James Gang lacked in pinup quality they made up for in talent, but you know how far that will get you in rock.

The downside of the power-trio boom was that it ushered in an era of exceedingly bland band names: West, Bruce & Laing; Beck, Bogert & Appice—suddenly it was cool to sound like a law firm. You even had power duos (see: Whitford-St. Holmes). Finally, the point became to choose the most boring name you could possibly think of. What else could explain the success of Hamilton, Joe Frank & Reynolds? Even the Hollies succumbed, calling a late-period LP *Clarke, Hicks, Sylvester, Calvert and Elliot*. But the sad fact is, five years post-"He Ain't Heavy (He's My Brother)," the Hollies were just looking to see if anything would stick—be it a couple of disco tracks, or what looks on paper like a nakedly bad move: an attempt at hard rock called "48 Hour Parole." See? There was a point to that digression.

As the Hollies struggled to remain relevant, the Clash began their rise to the status of "the only band that matters," releasing their self-titled debut in early 1977. Because it was excised, along with three other songs, from the American release of *The Clash*,

"48 Hours" may not be familiar to stateside listeners. The move pissed off the Clash, but I'd be surprised if many people grumbled about a release that included "Complete Control," "(White Man) in Hammersmith Palais," and "I Fought the Law" instead of "48 Hours," "Protex Blue," and "Deny." Nevertheless, "48 Hours" comes from the band's brief youth, before dub and geopolitics twisted the music into new shapes. Mick Jones said he dashed off the whole thing—music and lyrics—in half an hour. And it shows. Originally titled "48 Thrills," which served as the name for a fabled punk fanzine, this is a no-frills rocker that's distinguished by a couple of rarities: for one, Joe Strummer takes a guitar solo. And it's a celebration of kicks by a band that conspicuously had headier topics on its mind.

THE VERDICT

The anger and iconoclasm of punk may have been the prevailing mood in 1977, but roughly five years earlier, England had been seized by glam, a movement that was about fun and kicks and little else. As glam took flight, so did the career of the leather-cat-suit-clad Suzi Quatro, a Michigan native with a take-no-prisoners delivery beyond her years. At 21, she was the first female bass player to front a major band. Before becoming an international mega-seller, Quatro found success in England, performing songs supplied by the savvy team of Chapman and Chinn, who understood the power of blunt, stubborn hooks. Still, they were cocky enough to work in a notably darker vein than the twinkly-eyed sex romps of T. Rex or the rhinestone-encrusted theatrics of Gary Glitter.

Quatro followed her British No. 1 single "Can the Can" in 1973 with "48 Crash," a merciless ode to male menopause. "You've got the kind of a mind of a juvenile Romeo-o," she taunts. "And you're so blind you could find that your motor ain't ready to go-o-o." Making the most of a couple of overdriven hard-rock chords, vaguely tribal sounding drums, and a damnably simple but catchy chorus hook, "48 Crash" is a model of brute efficiency. The double meaning of "lightning flash" was no doubt lost on the kids, and what Suzi meant by "a silk sash bash" was anyone's guess. But who cared? Visceral, crunchy, and catchy was enough to land you at the top of the charts in that spangled era.

Suzi Quatro's stature as a prototypical tough female rocker cannot be overstated. A clear line runs from Ms. Quatro to the Runaways to L7 to Hole to Sleater-Kinney to Sky Ferreira. Still an active performer, she celebrated 50 years of hard rocking in 2014 and is determined to rock on: "I will retire when I go onstage, shake my ass, and there is silence."

A SONG TITLE	AN ALBUM	A BAND	A LYRIC
"48 Responses to Polymorphia" – Jonny Greenwood (2011)	*48:13* – Kasabian (2014)	Warhammer 48K – Missouri, psych-drone (2000s—2010s)	"It was forty-eight hours till Lonnie came around" – Steely Dan, "The Boston Rag" (1973)

> "In a cavern, in a canyon;
> Excavating for a mine;
> Dwelt a miner, forty-niner;
> And his daughter Clementine…"

About a century before the Joe Montana era, "My Darling Clementine" made "forty-niner" a household word.[1] The traditional, oft-covered "The Days of '49" is also rooted in the California Gold Rush; it recounts "a few hard cases"—men who met their fate "in the days of old/when we dug out the gold/in the days of '49." Bob Dylan's version comes from his much-maligned *Self Portrait*—a double LP that included inferior versions of his own songs and seemingly tongue-in-cheek Paul Simon and Gordon Lightfoot covers. In 2013, when Dylan re-released his putative nadir work, reviewers were keen to trot out Greil Marcus's notorious pan of the record (which began, "What is this shit?"). In the final analysis, *Self Portrait*, as originally released, is neither an outrage nor a misunderstood classic. Call it a grab bag with an unusually low ratio of hits to misses for a guy like Dylan. "Days of 49" is clearly one of the hits. Originally an upbeat folk number, Dylan does it as a rocking cowboy song in the rustic vein he'd been mining, both musically and sartorially, since *John Wesley Harding* in 1967.

"49 Bye-Byes," the final song on Crosby, Stills & Nash's first record and one of four by Stills, sports plenty of the trio's trademark harmonies, but it comes up wanting next to the heights of "Suite: Judy Blue Eyes" and others. Still, you can forgive S for not hitting all of them out of the park: While C and N were solely lending their golden throats to the enterprise, Stills was doing the lion's share, playing every instrument but drums and singing his ass off. As on virtually every song he's ever recorded, "49 Bye-Byes" finds Stills—who would soon embrace a look based entirely on jeans and sports jerseys—singing about his "lay-day." But this time, his line of seduction falls short of his best. "Steady girl, be my world" is no "Love the one you're with."

Technology has ensured that the obscure are faring much better than they used to, yet Nick Nicely remains obscure. The Londoner released a few singles in the early 1980s and then disappeared, leaving people like XTC's Andy Partridge in awe. Charged with coming up with a B-side for the beguiling "Hilly Fields," his labored-over 1981 masterpiece single, Nicely and his band dashed off the B-side in a day. "49 Cigars" sounds pretty spectacular, although the obvious sonic touchstones—in this case "Tomorrow Never Knows" and Syd Barrett-era Pink Floyd—were incorporated far more stealthily on the A-side. Nicely eventually resurfaced two decades later, giving his first-ever live performance in 2008 with his friends and fans the Bevis Frond.

---

1 State Route 49, which passes through many a historic California mining town, inspired songs called "Highway 49" by Big Joe Williams and Howlin' Wolf.

In a usage that one has to applaud for its stubborn mathematical sense, even as one decries the singer's excessive reasonableness, "Forty-Nine, Fifty-One," by country singer Hank Locklin, employs 49 specifically because it alone signifies the ratio of effort one man is willing to put into his relationship ("If you'll try forty-nine percent then I'll try fifty-one"). Of course, by the time the kicker comes around—he's willing to try fifty-two percent if she'll try forty-eight—you begin to suspect that old Hank is headed down a slippery slope.

In *The Crying of Lot 49*—a paragon of postmodernism (i.e., a book I don't really understand) by Thomas Pynchon—cultural references and historical digressions abound. The heroine of this short novel, Oedipa Maas, seeks to discover how she fits into the mysterious death, and life, of her ex, one Pierce Inverarity. It's not giving away a major plot point to mention that Lot 49 refers to a set of rare postage stamps up for auction. The book's tendrils have infiltrated indie rock at various junctures: At least three bands have called themselves Lot 49, and the addressee on a letter to Radiohead's merchandising arm is W.A.S.T.E. (an acronymic reference to the slogan of the book's Tristero organization). Most appropriate for discussion here is "Lot 49," a hyper blast of jangle by the wonderful English songwriter known as The Jazz Butcher. In its unforced shagginess and deadpan glee, the song speaks to a less fettered time in the world of indie music.

THE VERDICT

It's that biting Telecaster lick that sets the groove for the James Gang's "Funk #49" and makes it memorable. There's a middle section—with jungle noises and mucho cowbell that sounds flown in from a Kool & the Gang song—but it doesn't linger in the mind. What you remember is the force of those Joe Walsh chords and how they mesh with Tom Kriss's limber bass line and the, yes, seriously funky drumming of Jimmy Fox. You remember lines like, "Sleep all day/out all night/I know where you're going," but not necessarily the message itself. On paper, it's a warning against self-abuse, but when Walsh delivers those lines in his crooked croon, above that hot-asphalt riff, it feels more like a tribute to the very things the song ostensibly advocates against. Deep in my heart, I'm sure the people cranking "Funk #49" at all-night parties in the '70s were too busy shaking their hip-hugger-encased booties to feel scolded.

A SONG TITLE	AN ALBUM	A BAND	A LYRIC
"Gate 49" – Stiff Little Fingers (1981)	*49:00* – Paul Westerberg (2008)	49 Americans – London, art rock (1980s)	"Wake of the flood/ laughing water, forty-nine" – Grateful Dead, "Here Comes Sunshine" (1973)

50

"**50** Ways to Leave Your Lover," a smash hit from the golden era of solo Paul Simon, is the 500-pound gorilla in the world of #50 songs. And the song transcends mere popularity or sales; it is—to use a word I'm surprised Simon never used in a song—ubiquitous. It's been parodied ("50 Ways to Love Your Liver," "50 Ways to Fool Your Mother") as well as incorporated into the vernacular (the pilot's announcement on a JFK runway from 2008: "Remember, there are fifty ways to leave your lover, but only eight ways out of this aircraft").

Problem is, I have never been a fan of the song. I was overexposed to it at a tender age via a northern New Jersey A.M. radio station called WWDJ ("Ninety-seven, *DEE-JAY*"), and I still recall lurching across my bedroom to swipe at the radio dial in the nanosecond it took my synapses to recognize that distinctive military-snare opening. I admit it would be a bit churlish to sidestep a classic merely because it brings me back to the rainy days and Mondays of my youth, but something far beyond personal antipathy is at work here. "50 Ways" is dishonest song. Now wait—lest you think I'm about to hurl accusations of cultural piracy (the *Graceland* kind) at the man, I assure you it's nothing like that. This is purely about mathematics. To wit:

Slip out the back, Jack (Way$_1$)
Make a new plan, Stan (Way$_2$)
You don't need to be coy, Roy (A) (A=Advice)
Just get yourself free (A)
Hop on the bus, Gus (Way$_3$)
You don't need to discuss much (A)
Just drop off the key, Lee (Way$_4$)
And get yourself free (A)

In case it's not obvious, the point is that the man talks about 50 and doesn't even get into double digits. Now, I'm not saying he need mention all 50, like Sufjan Stevens does in "The 50 States Song." I would simply hold him up to the Shirley Ellis Standard. To wit, in "The Name Game," by going from "Shirley Shirley bo Birly" to "Arnold Arnold bo Barnold" before doing a little trick with Nick, she makes a pretty solid case for her claim that "there isn't any name that you can't rhyme." Simon, on the other hand, never does this for his 50-ways claim. The craft behind "50 Ways" is impeccable. But it's a Paul Simon song. Of course it's fucking well crafted. So I'll repeat myself (at the risk of being crude): a fatal lack of plausibility is this song's Achilles' heel.

Kate Bush's "50 Words For Snow" would seem to have some plausibility issues of its own. After all, the notion that the so-called Eskimo language contains a panoply of snow synonyms amounts to a collective cultural misunderstanding. But Kate seizes upon this faulty premise and makes it true—only in English. In fact, over a jungle thump and wintry synths, guest intoner Stephen

Fry's recitation makes a thrilling kind of sense, from "spangladasha" (#15) to "sorbet deluge" (#34) to the final [spoiler alert] … "snow." Take that, Paul Simon!

The Burnt Vegetables gave me my first taste of Frank Zappa. I was in junior high; they were a few grades ahead of me. With a name copped from Frank's "Call Any Vegetable," the band played at sporadic backyard parties. The set list featured originals along with Zappa's goofiest—"Take Your Clothes Off When You Dance" and "What's the Ugliest Part of Your Body?"—stuff even non-Zappa freaks could dig on. The Vegetables, God love 'em, would never have played a song like "Fifty Fifty." It requires some seriously sick chops, a migraine-inducing vocal, and a two-minute Jean-Luc Ponty violin solo, all far beyond the capabilities of even the most legendary unsung garage band.

THE VERDICT

Besides both being written by pint-sized performers, PJ Harvey's "50ft Queenie" and Paul Simon's "50 Ways to Leave Your Lover" have little in common. Polly Jean spits out the words at the top of her vocal range; Paul never breaks a sweat. "Queenie" has the stark, abrasive production of Steve Albini; "50 Ways" has the lush, detailed sound of mid-'70s album-oriented radio. Despite the tricky time signature and abrupt shifts in volume that make it radio-unfriendly in the extreme, "50ft Queenie" was a hit in the U.K. But it was just too unrelenting for mainstream U.S. radio. The song did manage to make its presence known stateside though, with a little help from MTV.

> **Butt-Head:** Beavis, the name of this song is "50 Foot Queenie."
> **Beavis:** Yeah, I'd like a 50-foot queenie.
> **Butt-Head:** I'd like a 50-foot weenie.

Weenie or queenie, the two cartoon cretins would back me up that "50ft Queenie" quite simply kicks ass. When the opening guitar figure—a swamp blues lick with its tail tied in a knot—gives way to mountains of guitar and Ms. Harvey starts to unleash, you almost want to run for cover. This enraged goddess has some choice words for the overburdened lothario of Simon's song: "You bend over/Casanova//No sweat/I'm clean// Nothing can touch me."

A SONG TITLE	AN ALBUM	A BAND	A LYRIC
"Fifty-Fifty Clown" – Cocteau Twins (1990)	*50 St. Catherine's Drive* – Robin Gibb (2014)	50 Cent – New York, hip-hop (2000s)	"Forget that I'm 50/ cause you just got paid" – David Bowie, "Cracked Actor" (1975)

I know what you're thinking. Why zero? Why now? After all, zero comes before 1, and we've been proceeding in numerical order. The fact is, it makes intuitive sense to begin at 1, and yet zero songs are plentiful enough to demand a proper airing. So here, at this halfway vantage between 1 and 99, seems the right moment to pause and consider nothingness.

The big goose-egg symbol we use to signify zero is a visual depiction of absence, and absence is rarely seen in a positive light in human affairs. Zero connotes both bereftness and extreme cold. By combining these two conditions—having no possessions and being at the mercy of the elements—you enter into a rare zone of suffering, in the spirit of King Lear or the protagonist in Sonny Boy Williamson's "Nine Below Zero."

Calling someone a 10 is high praise, but when Karen O looses "You're a zero" on a fame junkie in the Yeah Yeah Yeahs' spectacular "Zero," she sure isn't flirting. Equally biting is Public Enemy's "She Watch Channel Zero," in which Chuck D and Flavor Flav spit vitriol at a superficial woman who's watching all the wrong shows—the 1985 equivalent of the blank-eyed cellphone zombie. And when Billy Corgan proclaims, "I'm your lover, I'm your zero" in the Smashing Pumpkins song "Zero," it is not zero in the "I make multiplication easy" way, but rather zero in the "God is empty just like me" way. Less histrionically, Joan Armatrading poignantly captures the dizzying descent from "feeling like you're number one" to the hellish depths of love gone away in one of her signature songs, "Down to Zero."

The stage name of the great comedian Zero Mostel, who once hosted a TV show called *Channel Zero*, is believed to refer to the actor having arisen from nothing. And the elusive quantity's reputation has not been enhanced by its connection with a certain bucktoothed doofus in Mort Walker's Beetle Bailey comic strip, which first appeared in 1950 and likely helped cement the idea that Zero as a name signified an underachiever. In Springsteen's early ditty "Terry and Zero Blind," for example, Zero is the leader of a street gang called the Pythons, and "Terry's daddy understood that this Zero was no good." Of course Terry's daddy understood. He knew damn well that with a name like Zero, dude was not headed for a Ph.D. In "Zero From Outer Space"—a full-tilt swamp-rocker from Tom Petty & the Heartbreakers' underrated *She's the One* soundtrack—the alienated protagonist doesn't sound particularly bright either. Still, the recent preponderance of variously talented video-game characters named Zero shows the name moving beyond its ne'er-do-well connotations.

Books about numbers are full of references to "less than zero," but it took Elvis Costello to popularize the phrase in a nonmathematical sense. His 1977 song excoriating the British fascist Oswald Mosley produced headaches for the 23-year-old Costello when he cut short his performance of it on *SNL* and launched into "Radio Radio" instead. The stunt got him banned from the show for many years, but word was out. A Warsaw quintet called Lady Pank had a major hit in Poland with "Mniej niż Zero" in 1983. Two years later, when Bret Easton Ellis's bestseller made the phrase a generational signifier, Costello was still banned from the show for *not* singing the song that inspired Ellis's book title.[1] To be sure, Costello's song is a great one, and might be the best known zero song out there, but he's decidedly singing about something *not* zero, and that is not what we're looking for here.

It takes some enlightened thinking to see beyond zero's shortcomings. On the upside, zero enables writers to suggest a pithy juxtaposition of absence and substance, as in the Pop Group's protagonist "with zero reasons for living" in "She's Beyond Good and Evil," and "the zero times she calls me back" which refine the soul of Scott Miller in "Sodium Laureth Sulfate." Seems as if no one has opted to use the ordinal "zeroth" yet, but there's time.

The Fixx, a London-based synth-pop duo not generally acknowledged for paradigm shifting, bucked a couple of trends with "Saved By Zero," which is several cuts above the average '80s synth-pop radio hit. For one, this high-charting single extols rather than impugns the quantity in question. It even has a Buddhist subtext. Compared with Elton John's "Too Low For Zero," which was released one month later, the Fixx ditty is pure gold.

Along with the Fixx, the Internet made its official debut in 1983, heralding the digital age, but it took some time for the idea of ones and zeroes to settle into common parlance. Over a gritty thump that signals survival and defiance, the late-period Mekons song "Ones and Zeroes" mourns the human disconnection wrought by our hyper-connected world. A much-earlier Jesus Jones song of the same name, which sounds about as futuristic as the talking calculator from Pet Shop Boys' "Two Divided By Zero," underwent a pair of radical "reconstructions" by Aphex Twin on his *26 Mixes for Cash*, an intervention that revealed the dystopian coldness of those barren digits.

The Meat Puppets sincerely contemplate math and nothingness in "The Mighty Zero" ("It's just a number/You can count it on your hand/But I still don't seem to get it/ There's nothing to understand"). Similarly lysergic, "Zero the Hero" by the '70s-era prog rock collective Gong is a 10-minute suite that introduces several mythological entities who would populate the group's subsequent records. "Zero the Hero" is also a song by the Ozzy-less Black Sabbath, from 1983's (there's that year again!) roundly reviled *Born Again*. (This tour featured a wrong-size Stonehenge set that was immortalized in *Spinal Tap* a few years later.) With the ubiquity of the zero/hero rhyme, credit for creativity goes to Billy Bragg for "I look like Robert DeNiro/I drive a Mitsubishi Zero" in the still-stirring "Sexuality."

---

1 A year after Easton Ellis went global, the Stones were at their nadir: "Back to Zero" is a weak track from 1986's *Dirty Work*. Much finer, and similarly mining zero for its "ground zero" nuclear implications, is the excellent "Oh Zero" by Clocker Redbury & Dusty Slosinger (aka Jimmy Smith of the Gourds).

Billy Preston also had a way with a phrase. When he told a pining Stephen Stills to "love the one you're with," it inspired the newly solo rocker to write himself a huge hit. Wisely, Billy kept "nothing from nothing leaves nothing" for himself. If you do the math, the 1972 Top 10 hit based on this phrase would be a zero song to rival Nilsson's "One." But there is no Z-word in the title. Nor is there one in the lyrics of Bob Dylan's "Love Minus Zero/No Limit," which is more about love than about zero anyhow. And surely that won't do. Any song worthy of the zero crown must invoke zero qua zero in some way.

The minds behind *Schoolhouse Rock!* deliver a nuanced portrait of the elusive quantity on "My Hero, Zero," shifting from a light pop workout to sly jazz riffing while subtly advocating for doing your own thing. Another gem from the show's master composer, Bob Dorough, "My Hero Zero" even touches on infinity in charming, kid-friendly terms: "No one ever gets there, but you can try."

THE VERDICT

The Idaho-based trio Built to Spill were at the forefront of indie rock's late-'90s re-appropriation of guitar-based rock music, stripping away its heroic poses and melodrama and offering a seriousness of intent and the vulnerable perspective of leader Doug Martsch. Often described as a guitar hero, Martsch harnessed his early prog tendencies to forge a style wrought from recognizable facets of Neil Young, J. Mascis, Johnny Marr, and others, but bent them all to his own stubborn will.

"Carry the Zero," from the major-label debut *Keep It Like a Secret*, hits all the right notes. It's determinedly concerned with the unique character of zero, and it has the emotional heft required to suit a multifaceted quantity. The math in this intimate first-person plaint is rather ingenious. Martsch reaches the title phrase ("Didn't add up/forgot to carry a zero") in the first verse and leaves it behind, but the language is telling. The singer is all too human, supposing this relationship's failure was just a matter of missed details and not seeing the futility of carrying the zero—which has no effect in equations or interpersonal affairs.

Rooted in the sweeping melancholy of the major seventh chord, the chord pattern of "Carry the Zero" has a yearning, distinctly Smiths-ian flavor, which adds a sense of pathos. Two people together do make an equation, and sometimes it just does not compute: "You have become a fraction of the sum," Martsch sings in his almost pure tenor. But all that mathematical hoo-ha goes right out the window at 3:55, as the song flies away in an extended outro of visceral guitar lines that entwine and bend until a final, blissful fade.

A SONG TITLE	AN ALBUM	A BAND	A LYRIC
"Zero Refills" – The Pernice Brothers (2006)	*Revolution Number Zero* – Brian Jonestown Massacre (2013)	Zero 7 – London, electronica (1997–2010s)	"He acts the hero/we paint a zero on his hand" – Sonic Youth, "Teenage Riot" (1988)

51

The top-secret military testing ground in the Nevada desert known as Area 51 holds a place in the collective imagination as a hotbed of extraterrestrial life. Will Smith ends up at the site—which the U.S. government only admitted existed in 2003—after crash-landing in *Independence Day*, and the scores of songs with "Area 51" in their titles attest to its inspirational qualities. The post-Gram Parsons Flying Burrito Brothers, the Charlatans UK, the Yeah Yeah Yeahs, Yngwie Malmsteen, and Graham Parker have all mined Area 51 for subject matter. So have stalwart Portland thrash-mongers Dead Moon and Paddle Cell, purveyors of Teutonic psychobilly.

"Dick Butkus #51" is Dillinger Four's ode to the legendary Chicago Bears defensive end who once said, "When I played pro football, I never set out to hurt anyone deliberately—unless it was, you know, important, like a league game or something." "51%" is a dreamy morsel of muted optimism from Mark Sandman, the leader, singer, and sax player of Morphine, who died after collapsing onstage during a performance in Rome in 1999. Sandman's husky whisper rides a cool wave of sax, two-string bass, and plucked slide guitar, and the sound is plain gorgeous. "51 Phantom" by the North Mississippi All-Stars, emanating from a nearby swamp, sounds just right next to the Sandman's heavenly drone.

Nearly half of the United States has a Highway 51, so a mess of Highway 51 songs is to be expected. On *Bob Dylan*, the toast of Hibbing, Minn., covered "Highway 51 Blues" by Curtis Jones, in the urgent Woody Guthrie style that marked his early work. The Jones version makes clear that the highway in question is U.S. 51, which runs from Wisconsin to New Orleans, but John Lee Hooker doesn't pay much mind to the actual route on "Goin' on Highway 51"—he's too busy lamenting the loss of Miss Fannie Mae, who wouldn't even shake his hand when she left. All she said was, "Someday I will meet you when your troubles are like mine." Now that's a good highway song.

Around the time of the 1966 premiere of *Blow-Up*, the film's director, Michelangelo Antonioni, set out to make the definitive counterculture movie. Two years later he began filming *Zabriskie Point*, featuring trippy incidental music by Jerry Garcia, an uncredited Harrison Ford as a student agitator, and a tagline that sounds like it was coined by Matthew McConaughey's character in *Dazed and Confused*: "Zabriskie Point. How you get there depends on where you're at." The film's orgy scene was set in Death Valley, and the incendiary climax was soundtracked by "Come in Number 51, Your Time Is Up," by Pink Floyd. This slow-building freak-out ended up on the band's half-live, half-studio *Ummagumma* in 1971, as "Careful With That Axe, Eugene," but the version heard in the film is in a different key and lacks the all-important whispered warning of the title. (Eugene, by the way, was definitely playing with a deck of 51, as the Statler Bros put it in "Flowers on the Wall.")

I bet Aimee Mann, whom reviewers never fail to describe as "acerbic," would appreciate the irony of segueing from "Fifty Years After the Fair"—one of her bounciest songs—to the lethal comedown of "High on Sunday 51." In the former song, she proclaims, "We'll get it right, I swear," while "Sunday" takes the grim view of mankind that is more typical of Ms. Mann. Just as Elvis Costello is apt to depict relationships in military terms, Ms. Mann has a preferred metaphor: addiction. In a few strokes, the refrain of "High on Sunday 51" conjures the fool's bargain made by the enabler.

"51st Anniversary" is an obscure Jimi Hendrix song that merits serious consideration because it's a Jimi Hendrix song. Reissued on the 1994 CD re-release of his monumental *Are You Experienced*, the song was the B-side of "Hey Joe" and did not make it onto the original U.K. or American vinyl releases. Truth be told, it's not hard to see why this was slapped on the back of a single and forgotten about; it just isn't that good. In fact, "51st Anniversary" has as little going for it as any Hendrix song I can think of. The chord progression feels pedestrian, the lyrics scream first-draft, and the spoken-word section sounds like practice for something Jimi did a hundred times better on "If 6 Was 9." There's a reason this book isn't called *51st Anniversary and Other Assorted Number Songs*.

PALATE-CLEANSING FACT #51
Brian Wilson wore #51 during his stint as quarterback of the Hawthorne High football team.

THE VERDICT

I was lucky to have an older brother who didn't believe in buying an album or two at a time; he saved up and bought them in stacks. *The Ghost of Cain* by New Model Army came from one of Jonny's stacks. After playing "51st State" for me (it was already a hit in the band's native Britain), he must have caught something in my crazed eyes that said he would never love this record half as much as I already did, and he gave it to me. I still cherish the LP, if only because it brings me back to the days when I could love a record so irrationally. The black-and-white photo on the back cover, showing the three band members glowering under ominous skies, evokes not even a shred of awe anymore. The man in the middle—the one with the most scornful expression—went by the nom de rock Slade the Leveller. At the time, that made a big impression on me. What a guy in his 20s could dig about New Model Army (named after Oliver Cromwell's antiroyalist militia) isn't hard to determine: The music was dark, precise, unforgiving—and catchy. They traded in politically charged anthems, with lyrics as much spat-out as sung, and their lack of a sense of humor was not something I viewed as detrimental.

"51st State" (actually a cover of a song by a band called the Shakes) was NMA's biggest British hit. Abetted by a rousing soccer-chant chorus, the song takes aim at Yankee imperialism and pulls no punches: "We're W.A.S.P.s/proud American sons/we know how to clean our teeth/ and how to strip down a gun." Members of the group have had difficulty obtaining visas to travel overseas ever since.[1] Personally, I never had a problem singing along to, sometimes even pounding my chest in sympathy with, the triumphant anti-American chorus. I was no W.A.S.P., but I was an American, and I dug the fury that New Model Army hurled our way, in much the same way as a baby monkey prefers being beaten by its mechanical mother to being ignored by her.

---

1 New Model Army did not limit its confrontational stance to the world's great powers; in the mid-'80s they were known to perform clad in T-shirts that declared "ONLY STUPID BASTARDS USE HEROIN."

Though they never established a substantial beachhead on U.S. soil, New Model Army maintains an extraordinarily devoted fan base in the U.K. Main dude Justin Sullivan retired Slade the Leveller long ago, but he continues to front various incarnations of NMA and remains committed to the pursuit of global justice. If there's a lesson to be learned here in Chapter 51, it's that the worst song in a great man's catalog is no match for a really good band's best. And there's some justice in that.

A SONG TITLE	AN ALBUM	A BAND	A LYRIC
"51-7" – Camper Van Beethoven (2004)	*51* – Kool A.D. (2012)	Spy 51 – London, alt-rock (2000s)	"The pope owns 51 percent of General Motors" – George Harrison, "Awaiting On You All" (1970)

52

Mathematically, 52 is an "untouchable" number—meaning it's never the sum of the proper divisors of any other number—and maybe this fact has some bearing on the demonstrable scarcity of 52 in the world of song. A deck of cards, the number of weeks in a year, these are the greatest hits of 52. So wouldn't it make sense that there'd be a gambler's lament called "52 Pickup," or some old chestnut with a refrain that talked about "loving you 52 weeks of the year"? There is indeed a handful of "52 Pickup" songs, but I'll be damned if any of them are notable. Certainly none can lay claim to being the musical equivalent of Elmore Leonard's *52 Pickup*, a taut crime thriller whose screen version starred Roy Scheider, Ann-Margret, and former Prince protégé Vanity. With its wafting clouds of gulping keyboards and Sarah Cracknell's cotton candy vocals, Saint Etienne's marimba-dotted "52 Pilot" would make a fine soundtrack for that film's party sequence, which featured XXX screen legends Ron Jeremy and Amber Lynn.

Robyn Hitchcock, whose principal touchstones are Dylan, Barrett, the Byrds, and Lewis Carroll, begins "Fifty-Two Stations" with two lines that positively thrum with information. "There's fifty-two stations on the northern line/None of them is yours, one of them is mine." In a single couplet, we know the singer is a spurned man, an obsessive type who knows too much about train schedules and rides the Tube lamenting lost love. He seems resigned and wistful at first, but eventually sorrow turns to anger. Now he's haunted by her memory, wanted for assault (if the police are doing their job), riding the Northern Line (the black line on the color-coded London Tube map), a shadow of his former self and a menace to his fellow riders. For a man whose catalog includes songs like "Veins of the Queen," "The Man With the Lightbulb Head," and "Sandra's Having Her Brain Out," a song like "Fifty-Two Stations"[1] is pretty straightforward stuff. No insects, no Egyptian cream, no one having her brain out—just a desperate man who's smitten enough to see the face of the woman who done him wrong, whenever the train stops.

THE VERDICT

As the story goes, one night in 1976 after collectively sharing a mystical libation at a Chinese restaurant, five friends from Athens, Georgia, jammed for the first time and dubbed themselves the B-52's. Apparently the name had come to drummer Keith Strickland in a dream. The term traditionally refers to a strategic bomber that figured prominently in the Cold War and *Dr. Strangelove*, but it

---

[1] There are actually 50 stations on the Northern Line, the London Underground's second busiest, so either Hitch was channeling a ghost (entirely possible in his case) or he needed an extra syllable. "I was never intentionally obscure," Hitchcock once said. "It's just that everything seemed to me so confusing that my songs always seemed very fragmented 'cause that's how I perceive things."

had an additional meaning as a slang expression for the towering beehive hairdos, better known as bouffants, favored by vocalists Cindy Wilson and Kate Pierson.

"52 Girls" is immediately arresting, a precise calibration of Keith's unignorable drum beats, Ricky Wilson's sinewy guitar riffs, and Cindy and Kate's piercing vocals. (The temporary absence of Fred Schneider's *Sprechgesang* does not feel like any kind of loss.) The song seems to be a celebration of girls, which is not unusual—only it's sung by girls, and that is unusual, especially in 1977. And it's not about the typical girls of song—heartbreakers, teases, impossible dreams, etc. This is about the girls who aren't clichés: "Tina, Louise and Hazel and Mavis." "Wanda and Janet and Ronnie and Reba." *These* are the true cool girls of the U.S.A. Thirty-eight years ago, when "52 Girls" came out as the B-side of "Rock Lobster," that was a bold statement, perhaps even quietly revolutionary. Yet the message, if I read it correctly, was not easy to decipher. You can listen 100 times and still not hear "Effie, Madge and Mabel and Biddie" as the song's opening line. No doubt the sheer elusiveness of Kate and Cindy's vocals, veering from pep-rally clarity to something bordering on pure sound, is part of the song's enduring appeal.

No one has ever managed to look or sound like the B-52's. When John Lennon first heard them, he was inspired to return to the recording studio after several years of musical inactivity. So arresting was the band's blend of influences (a wag at *People* magazine likened their sound to "the illegitimate offspring of George Jetson and the Shirelles") that in the hands of lesser mortals it would have come off as mere camp. Instead, the B-52's projected nothing less than total commitment. Surely the most unheralded element of the band's first two records is Ricky Wilson's distinctive guitar playing. The B-52's could have not have existed without Wilson's work on a four-stringed, custom-tuned Mosrite. Simply by removing the two middle strings and tuning the remaining pairs way down (down! down! down!), Wilson achieved a limber yet punk-toughened take on surf guitar that, even without a proper bass, made the music really swing.

A SONG TITLE	AN ALBUM	A BAND	A LYRIC
"52 Favorite Things" – The Unicorns (2003)	*52nd Street* – Billy Joel (1978)	Energy 52 – Germany, trance (1990s)	"I heard you on the wireless back in '52" – The Buggles, "Video Killed the Radio Star" (1979)

LAST WORDS:
"Love is two minutes, 52 seconds of squishing noises."—Johnny Rotten

53

I admit I haven't read *Car, Boy, Girl*, the 1961 book on which *The Love Bug* was based, so I cannot say for certain whether the automotive protagonist of Gordon Buford's novel wore No. 53, or even if he was named Herbie. Not so surprisingly, *Car, Boy, Girl* is out of print, so please forgive me for not tracking down a copy. What's important is that whoever came up with the numeral for the cuddly Volkswagen Beetle got it right. Fifty-three is a number devoid of flash. You encounter it in mundane places: the ass-end of your phone bill, a road sign, the entrance to your friend's apartment. It's therefore fitting that the few #53 songs out there all tend to incorporate 53 in mundane ways.

Seattle's Minus the Bear named itself after *B.J. and the Bear* (minus the bear, get it?), a cheesy '80s TV show in which freelance trucker B.J. McKay, his pet chimp, Bear, and a gaggle of lady truckers do battle with the nefarious Sheriff Lobo. (While B.J.'s truck lacked a name, and a mind, of its own, Herbie's influence was unmistakable in the way the orange-and-white Kenworth K-100 semi took right turns.) Deliberately or not, "Memphis & 53rd" from *Menos Del Oso* shares the same central credo as the theme music from *B.J. and the Bear*: keep on moving.

"53 Summer Street" is a 1968 single by Turquoise, whose members grew up in the same Muswell Hill neighborhood in North London as the Kinks' Ray and Dave Davies. As the Brood, the band recorded demos with Dave Davies in '66, and more demos a year later with Keith Moon and John Entwistle. Indeed, it's the Who's influence that's most evident on "53 Summer Street," with verses that recall "Pictures of Lily" and a touch of "I Can See for Miles" at the end of the chorus. But somehow this tale of a club owner who ends up in jail due to unnamed shady doings at 53 Summer Street never achieves liftoff.

The B-52s' "53 Miles West of Venus" has something of a "Planet Claire" flavor, but it feels like filler. Don't get me wrong; just because the only line in the song is the title, repeated endlessly, doesn't necessarily kill the party for me—I mean, "Why Don't We Do It in the Road" is pure genius—but this is nowhere in that league. (Amazing how, with the shifting of a single digit, this numerical thing turns champs into chumps.) Honorable mention goes to "Midwatch 1953" by the Fall from *The Unutterable*, which is like a Fall take on the death of HAL the computer in *2001: A Space Odyssey*, only instead of a slurred recital of "Daisy, daisy, give me your answer true," Mark E. Smith wheezes, "Who could foresee what happened in 1953?" with instrumental backing from what sounds like two unrelated songs and a damaged pinball machine.

THE VERDICT

Just as the B-52's own #52, their contemporaries the Ramones own #53. Both bands always flirted with a cartoon image but were at heart totally genuine about the music they made. Ramones songs dealt with harsher subject matter, of course, but most were leavened by deadpan humor and shout-along choruses. To say "53rd and Third" lacks the buoyancy of typical Ramones fare is to understate the case. Even the title, soullessly evoking the drab grid pattern of New York City streets and avenues, is short on glee. It simply imparts the location where the song's protagonist toils in the sex trade. And while the Ramones mostly purveyed a pile-driven version of bubblegum or girl-group pop, "53rd and Third" is more of a blunt onslaught. The chipper melody of a song like "Beat on the Brat" keeps it from feeling like a genuine celebration of assault, but in this squalid little tune, there is no subtext, no sweet spot, no place to hide. Dee Dee Ramone takes a rare solo vocal to deliver the crucial confession, "Then I took out my razor blade/then I did what God forbade." After a false ending at 1:48, there's a return to that infernal address, which Joey repeats, dissonant and unresolved, into the fadeout.

Many believe the narrative is based on actual events. When asked about it, Dee Dee was not cagey. "The song '53rd & 3rd' speaks for itself," he said. "Everything I write is autobiographical and very real. I can't write any other way."

A SONG TITLE	AN ALBUM	A BAND	A LYRIC
"Delta Flight 53" – John Fahey (2000)	*Yardbird DC-53* – Charlie Parker (1953)	53 Large Men – Texas, alt-rock (2000s)	"In the summer of '53 his mother brought him a sister"— Andrew Gold, "Lonely Boy" (1976)

ENDNOTE

53rd and Third is the name of a short-lived Scottish independent record label that Stephen Pastel founded in the mid-'80s. It was home to several of the so-called C86 bands, including the Shop Assistants and the Vaselines, whom Kurt Cobain famously loved, and several other un-Ramones-like bands.

54

If you look closely at Studio 54's white-on-black "54" logo, the 5—clearly the masculine of the two numerals—seems to be subtly humping the 4. And the salacious Disco Era connotations of 54 don't end there. Xenon, a popular but less remembered nightclub from the same period, took its name from the element whose atomic number just happens to be 54. Coincidence? Possibly, or perhaps it was a deliberate but subliminal nod toward the biggest thing out there, in the best tradition of the Sex Pistols inspiring the tweaked version of their name: Celibate Rifles. In any case, no song from that sozzled epoch actually uses a Studio 54-iented title, although several dance tracks from later decades do.

In 1972, five years before Studio 54 opened its hallowed doors, Harry Nilsson was at his commercial zenith. The Brooklyn native, who once caused Little Richard to exclaim, "My, you sing *good* for a white boy!" suddenly went from musician's musician (Paul McCartney called a then-unknown Nilsson his favorite American singer at a late-'60s press conference) to extremely successful recording artist. *Nilsson Schmilsson* had yielded three wonderfully diverse singles: "Without You"—a cover of a Badfinger song—was four minutes of Orbison-worthy melancholia; the lilting, utterly ridiculous "Coconut" had millions of people around the world humming "You put de lime in de coconut" in spite of themselves; and "Jump Into the Fire" was a thunderous slab of nerve-jangling rock 'n' roll that featured Nilsson's desperate, ragged vocal and an aggressive bass line played by Klaus Voorman.

Nilsson wasted no time in following up his commercial breakthrough with a complete about-face. *Son of Schmilsson* opened with "Take 54," a stomper whose refrain—"I sang my balls off for you baby!"—ensured that it would never be played on the radio. Today it still comes off as a pretty rude lyric; in the Nixon re-election year it was doubtless even more jarring. As the decade progressed, Nilsson explored a succession of noncommercial genres, including English music hall and the American songbook, thinning out his audience even more. That his best-known achievement of the late '70s was getting thrown out of the Troubadour in L.A. with John Lennon for heckling the Smothers Brothers says a lot about Harry's inauspicious career arc.

Like several other compositions on the ambiguously pronounced *Drukqs* collection by Richard "Aphex Twin" James, "54 Cymru Beats" has a Welsh name (Cymru is Welsh for Wales). But anyone hoping for something with a touch of the Welsh folk tradition—a fiddle perhaps—will be disappointed. Instead, "54 Cymru Beats" is a tangle of simultaneously caressing, scraping, whooshing, and pummeling sounds sewn together by an obsessive and inscrutable master's hand. But it's so un-Welsh sounding it may as well be Swedish, like the Dandelions, whose single "On the 54" was featured in a Volvo ad and certainly enhanced the clothes-shopping budgets of this snappy-dressing Stockholm five-some.

THE VERDICT

Standing high above this assortment of fifty-fouria is Toots & the Maytals' "54-46 Was My Number," an oft-anthologized, oft-covered classic of reggae featuring a divinely laid-back groove offset by the exhortations of Frederick "Toots" Hibbert and his spirited backup singers, Nathaniel "Jerry" Matthias and Henry "Raleigh" Gordon. He wrote it after serving an 18-month prison sentence—not for possession of ganja, as is commonly believed, but simply for showing up to bail out a friend. While inside, he became fully committed to his Rasta identity, and the singles he put out following his incarceration, including "Do the Reggay" and the immortal "Pressure Drop," have earned a well-deserved place in the pantheon.

"54-46 Was My Number" is a picture of life behind the walls, distilled down to the singular humiliation of having one's name replaced by a number (which Hibbert made up for the song). The repeated phrase "listen what I say" serves as rhythmic linchpin and seems to echo Ray Charles's "What'd I Say?", which also features stops and starts, call and response, and multiple release points. And like the Ray Charles song, "54-46" draws on the blues and gospel traditions, creating something joyous and danceable out of pain and injustice.

A SONG TITLE	AN ALBUM	A BAND	A LYRIC
"54 Chevy" – Michael Hurley (1984)	*54 Days at Sea* – Chris Bailey (1994)	2:54 – London, alt-rock (2010s)	"A '54 convertible too, light blue" – Eartha Kitt, "Santa Baby" (1953)

55

Charlton Heston, Ava Gardner, and David Niven starred in *55 Days at Peking*, a 1963 film about China's Boxer Rebellion. Sammy Hagar, an avid boxer in his youth, became known for rebellion with "I Can't Drive 55." I won't venture a guess as to how Charlton, Ava, and David would have fared, but it's a good thing Sammy wasn't born in Victorian England, where the Locomotive Act—the world's first speed limit—made it illegal to drive a car (known then as a "light locomotive") faster than about 10 mph. My guess is that Hagar, a longtime Patti Smith fan, (they jammed together when both were inducted into the Rock Hall of Fame in 2007) would have had to invent punk 100 years ahead of schedule just to express his outrage.

The Screaming Blue Messiahs also sang about the accursed speed limit in the late '80s. "55 the Law" starts off a straightforward rockabilly celebration of the open road, until the singer slips something in about "the wife and kids are dead"—an odd touch indeed. Led by the chrome-domed Bill Carter, the Messiahs had a punkish take on good ol' *Amurican* roots rock that earned them the blessing of David Bowie, but their street cred took a fatal dive after "I Wanna Be a Flintstone."

"5:55" is the bewitching title track to Charlotte Gainsbourg's first grown-up solo work. Co-written by the French duo Air and Jarvis Cocker of Pulp, the song is a lush and transporting blend of rolling piano chords, whispered vocals, and soaring strings. There are more than a few songs titled simply "55"—by Echoboy, San Antonio troubadour Jack Levitt, and even the Master Musicians of Jajouka, who were to Brian Jones what Ladysmith Black Mambazo was to Paul Simon. Seek out "Fifty Five" by Pink Industry, an eerie slice of synth-pop from a duo comprising former Frankie Goes to Hollywood bassist Ambrose Reynolds and former Big in Japan vocalist Jayne Casey.

To the youth of Boulder, Colorado, circa 1963, the Astronauts—Rich, Stormy, Bob, Dennis, and Jim—were the biggest thing around. "55 Bird," the band's pleasantly goofy tribute to a well-loved vintage of Ford Thunderbird, employs a vocal arrangement reminiscent of their contemporaries the Beach Boys, who had already transcended the surf music genre in a way that bands like the Astronauts and the Trashmen (proud sons of Minneapolis and the creators of the classic "Surfin' Bird") never would. "55 Bird" is a fun trifle, but the band's fever-charged instrumentals—powered by a twin rhythm guitar attack—were its strong suit. The Astronauts' lone chart success came in 1964 with a sizzler called "Baja" written by Lee Hazlewood, the famed producer/songwriter-for-hire. Hazlewood, who went on to record 20 idiosyncratic albums of his own (most of which went unappreciated until the end of his lifetime), had an enlightened rogue persona not far from that of Tom Waits. Hazlewood even recorded a Waits song on his 1973 LP *Poet, Fool or Bum*, which received the one-word review "Bum" upon its release. That year, while Hazlewood's grizzled, booze-soaked melancholia was getting no respect at all, his musical doppelganger over at Asylum Records turned out a grizzled, booze-soaked masterpiece that kick-started his career. Last time I checked, it was still going strong after four decades. No one said rock 'n' roll was fair.

THE VERDICT

"Ol'55" opens Tom Waits's proper debut, *Closing Time*, and it's hard to feel anything but instant kinship with this love song for a car. So lonely it aches, then soaring and full of hope, "Ol'55" introduced the world to a voice that one waggish writer said sounded like it was bathed in whiskey, hung in a smokehouse, and then run over.

When the world-weary Waits (who was only 24 at the time) turns to his bucket o' bolts, his auto is more than just a ticket out of a bad situation; it's refuge and salvation. It's the thing we hang onto to keep from falling into mortal despair, the thing with feathers. A year later, his label-mates the Eagles recorded their own version, liberally sweetened with West Coast harmonies, which is the version most people know. Ian Matthews and Sarah McLachlan covered it, too, some would say well, but the original's gravelly grace is something the voices of Henley, Frey, Matthews, and McLachlan are just too damned pretty to capture. The adverb "lickety-splicky," for example, should only be sung by people whose voices have been run over at least once.

A SONG TITLE	AN ALBUM	A BAND	A LYRIC
"Fifty Five Falls" – Marissa Nadler (2004)	*55* – Vibro Kings (2004)	Primer 55 – Kentucky, nu-metal (2000s)	"And now it's 55 years later/ We've had the romance of the century" – Magnetic Fields, "Papa Was a Rodeo" (1999)

POSTSCRIPT: Whether "Schfifty Five" by Group X is technically eligible to win the top spot is a question I will leave to the numerological sages on high. Thankfully, Tom Waits has made the question moot, but this strangely inflected rap goof by a Georgia band posing as an Arabian outfit has some kind of primitive magic to it.

David Klein

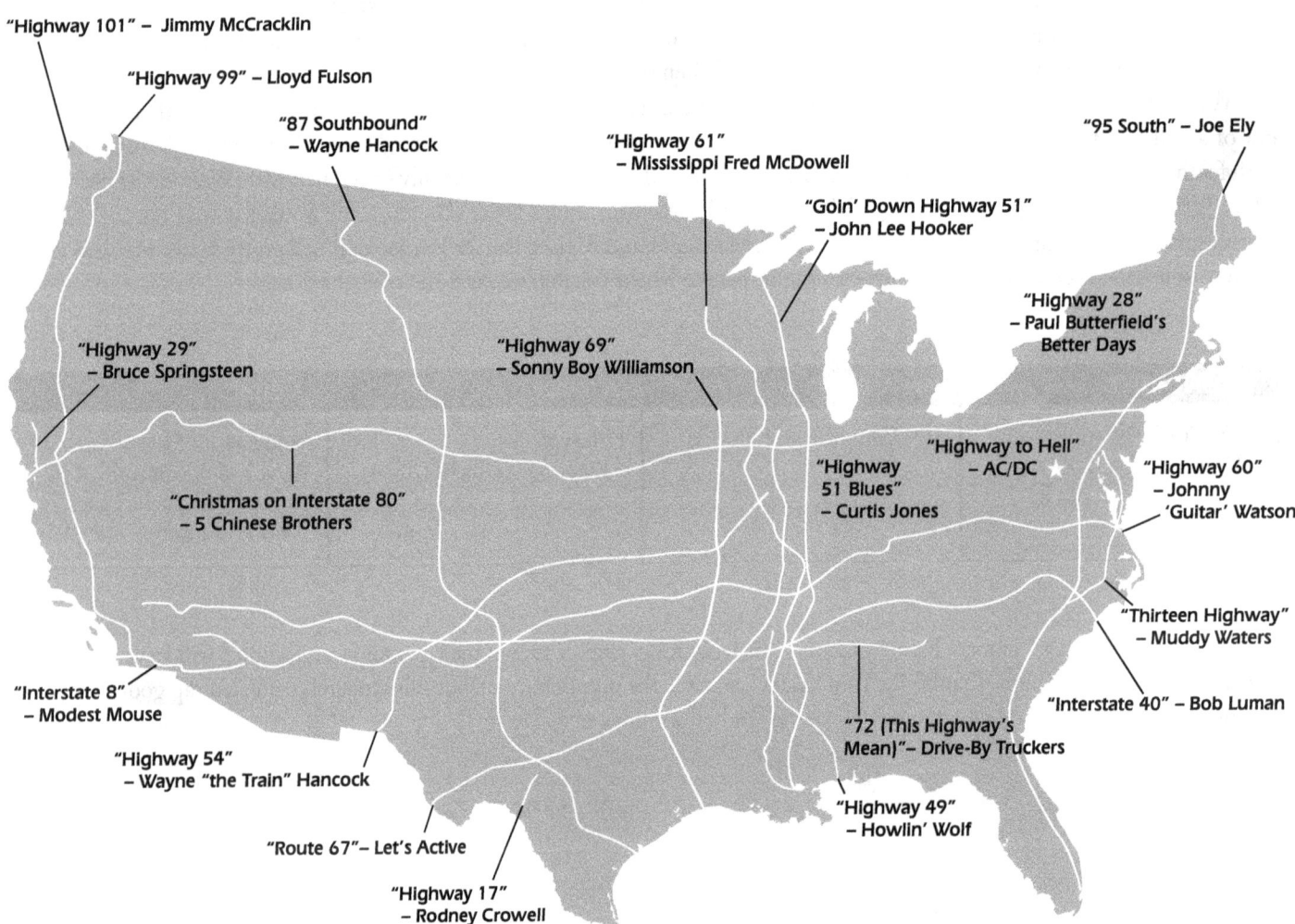

56

Fifty-six finds its way into some primo sepia-toned moments of the past century—Joe DiMaggio hit in 56 consecutive games, Shirley Temple had exactly 56 curls on her head, to name two—but these admittedly alluring phenomena did not inspire songwriters to render specific numerical homage. Thus, 56 is perhaps best known to rock enthusiasts not through a song title but by a brief but memorable walk-on part in a Bob Dylan song that begins, "Meet me in the morning/56th and Wabasha." No more eloquent mention of 56 exists in the annals of popular music: 56th and Wabasha, where Dylan dreams about meeting the lover whose absence torments him throughout *Blood on the Tracks*, is the pinnacle of explicit musical 56-ness. But should you be considering a pilgrimage to 56th and Wabasha as part of a mad quest to visit every place ever mentioned in a Dylan song, think again. The Wabasha Street near the University of St. Paul, where Dylan spent formative time, does not intersect with 56th.

Another, much huger thing that it pains me to note is that this admittedly rich discussion has focused on a song called "Meet Me in the Morning" and not "56th and Wabasha," which forces me to acknowledge that after a triple run of classic songs by classic artists (B-52's, Ramones, Tom Waits), numerical reality has slapped us upside the head, pointed an impudent finger at our chests, and said, "Are you *sure* about this? Is there really a good song, make that a damn good song, with 56 in its title?" The answer is yes, barely. On a technicality—but yes. (Be patient: we'll get to it.)

56 YEAHS
On R.E.M.'s Andy Kaufman tribute "Man on the Moon," Michael Stipe sings the word "yeah" 56 times. On Nirvana's yeah-fest "Lithium," Kurt Cobain sings 56 yeahs.

Several in the rarified world of 56-titled songs refer to 1956, a vital year in rock 'n' roll's infancy when the charts were clogged with Elvis, Buddy Holly, and Little Richard. "Nineteen Fifty-Six" by the Rascals—one of the premier singles bands of the '60s—is good fun, a rocked-up blues number that borrows perhaps a bit too much from "Kansas City." "Nineteen Fifty Six, Fifty Seven, Fifty Eight," a jaunty Bollywood rave-up celebrating the rush of progress, comes from a 1959 film called *Anari (The Naïve One)*. It features the distinctive vocal talents of one of the most celebrated Bollywood playback singers, Lata Mangeshkar, once alleged to be the world's most prolific recording artist and now acknowledged to be merely a fantastically prolific one, with many thousands of recordings to her name. Bollywood music is based on ragas and other traditional Indian structures, as well as a kitchen sink of influences; every one seems to make an appearance in "Nineteen Fifty Six." Oddly enough, it has a far stronger Eastern European flavor than "Budapest '56" by Paris Violence, a song about the infamous Soviet crackdown on Hungary, told via shouted French vocals and Ramones chords.

Orange County action-figure band Sugar Ray named "56 Hope Road" after Bob Marley's address in Kingston, Jamaica, while Goldenboy, a skate punk band from the west coast (of Norway), cite Paul Anka, Chuck Norris, and White Lion as influences; they sound a note of Weezerian power punk on "Fifty Six."

Bringing a jaunty ska beat is "Dub 56" by the Toasters, a long-running American revivalist institution whose members appreciate the sound of a good saxophone and would no doubt dig "Fifty-Six," a marvel of invention and technique by the legendary tenor sax man Johnny Griffin. "4:56 A.M." from Roger Waters's midlife-crisis-themed album, *The Pros and Cons of Hitchhiking*, is graced by plenty of Floydian sax, courtesy of David Sanborn. And "A Dip in the Lake: Ten Quicksteps, Sixty-two Waltzes, and Fifty-six Marches for Chicago and Vicinity" doesn't call for a saxophone, *but that doesn't mean that there isn't one*.

Conceived by John Cage in 1978, this theoretical work called for groups of people to visit hundreds of predetermined addresses in Chicago and "either listen to, perform at and/or make a recording of the sounds at those locations." So if a man happened to be playing saxophone at one of Cage's addresses, and one of the delegations opted to make a recording of him, you could say the work had a saxophone in it. But that's far too esoteric for me. At least with Australian black-metal exponents Spear of Longinus (named after the spear that pierced the side of Christ) and a song like "The Sine of Satan Is 56," you know damn well there's no saxophone, and you're glad for that certainty.

THE VERDICT

Certainty has been in short supply during my search for the ultimate #56 song. While I prefer to confer top honors on a title that uses the numeral in a deliberate or evocative way, sometimes that's just not possible. The song I've chosen, "Five Feet of Lovin' 56" by Gene Vincent and His Blue Caps, is identical to the original 1956 version of "Five Feet of Lovin'." (It was not unusual for Vincent to revisit songs from his back catalog, a practice that yielded a slew of alternate takes and alternate titles.) What really matters is that "Five Feet of Lovin' 56," by any name, shows off the talents of a singular, tragic figure in all his snarling rockabilly glory.

Born Vincent Eugene Craddock in rural Virginia, Gene Vincent came storming out of the gates in 1956 with "Be-Bop-A-Lula." Inspired by the Little Lulu comic book character, it's one of the defining songs of the early rock era. But Vincent never came close to the upper reaches of the charts again. Abandoned by American radio, he found favor and adulation overseas. But while on a 1960 tour of the U.K., he was in the horrific London taxi crash that killed his fellow young rocker Eddie Cochran. Vincent was permanently damaged. For the next 11 years, on various labels and amid numerous personal crises, he struggled to revive his career. In 1971, while visiting his father in California, Vincent died of complications related to a bleeding ulcer. He was 36.

As a preteen I learned about Gene Vincent from the hallucinatory *Rock Dreams*, by Belgian artist Guy Peellaert and pioneering music writer Nik Cohn, a book that distilled two decades of rock iconography and poured it into the folds of my fevered teenage brain. One haunting illustration showed a hunched, switchblade-clutching Gene Vincent, surly and defiant, cornered in an English pub and facing down a constable holding a badge. The accompanying passage is something

END NOTE
The unique taste of Jagermeister stems from its mix of 56 herbs.

If 6 Was 9

I've never been able to forget: "After he hurt his leg, Gene Vincent always performed in pain and the possibility of collapse, and he stood on stage without moving, leaning forward, with his bad leg half-bent in front of him. Sometimes he seemed quite desperate, and he would shudder and strain and shake himself like a maimed, black-leather animal, castrated by captivity."

A SONG TITLE	AN ALBUM	A BAND	A LYRIC
"56" – Mory Kanté (2000)	*Project 56* – Deadmau5 (2005)	Flatfoot 56 – Chicago, Celtic punk (2000s—2010s)	"I got a '56 Mercedes-Benz I had painted cocoa-brown" – David Johansen, "Funky But Chic" (1978)

57

In *Passenger 57*—a vehicle for Wesley Snipes—his character's tagline is "Always bet on black." H.J. Heinz had a much better slogan. When he adopted "57 Varieties" for his rapidly expanding foodstuffs company in 1892, Heinz gave 57 the kind of publicity you just can't buy. His choice of number had nothing to do with accuracy (the company's offerings were already well in excess of 57, thank you very much) and everything to do with catchiness.

There's both a nice ring and an uncanny aptness to 57, with its suggestion of overabundance that skirts outright hyperbole. Richard Thompson seems to invoke 57 in its Heinz-ian sense in "Valerie," a song about a frivolous temptress who spends her would-be suitor's money on "fifty-seven things she's never going to use." And it doesn't seem far-fetched to suggest that Bruce Springsteen, at least subconsciously, had ketchup on his mind when he wrote "57 Channels and Nothing On," an anti-TV diatribe from the generally underwhelming *Human Touch*.

"Class of '57'" by the Statler Brothers (none of whom were named Statler) is like a country version of Jim Carroll's "People Who Died," only instead of Carroll's New York-centric laundry list of fatality ("Judy jumped in front of a subway train/Eddie got slit in the jugular vein"), the Statlers give us, "Randy's on an insane ward/Mary's on welfare." Though not without its charms, the song lacks the wry humor of "Flowers on the Wall," the 1965 ditty used to such fine effect in *Pulp Fiction*.[1]

The Statlers, a Virginia quartet, began as the Kingsmen but were forced to rechristen themselves when the Kingsmen (of Portland, Ore.) became instant rock legends with their version of Richard Berry's "Louie Louie" in 1963. Fortunately, the Statler Tissue Company was around to provide inspiration.

THE VERDICT

Man Sized Action—Pat Woods, Tony Pucci, Kelly Linehan, Brian Paulson, and Tippy —was a major force in the vibrant early-'80s Minneapolis alternative rock scene. *The Five Story Garage* EP, released in 1984 on Bob Mould's Reflex Records, still sounds as fierce and unforgiving as the Minnesota winter, undiminished by time, new production techniques, or radical improvements in the science of rock 'n' roll. "Fifty-Seven" combines the loud-and-fast hardcore aesthetic with the corrosive guitar frequencies

---

1 It's playing in the car when Butch unexpectedly encounters Marcellus Wallace, right before their hellish fight lands them in the clutches of Zed.

of Hüsker Dü, led by the urgent keening of vocalist Pat Woods. According to Steve Albini's history of the band, the song's name derives from "its position on the MSA Master Song List of History and Achievement." Now that's a list I would love to see.

A SONG TITLE	AN ALBUM	A BAND	A LYRIC
"57" – Killdozer (1985)	*57 Octaves Below* – Air Formation (2005)	57 – South Africa alt-rock (1990s–2000s)	"Walking 'round the room singing/"Stormy Weather"/At 57 Mt. Pleasant St." – Crowded House, "Weather With You"

58

Even before I discovered that in Central American lore 58 signifies bad juju (something to do with 58 original sins), the number was already emitting perplexing vibes and wafting them my way. Fifty-eight (which is the sum of the first seven primes) presents a challenge to even the most seasoned seeker of numerically titled ditties. The fact that 58 is the name of the side project of Mötley Crüe bassist Nikki Sixx (their cover of "Alone Again (Naturally)" borders on a criminal act) supports my contention that the number is inherently flawed. Whether you agree or not, it's hard to argue that the offerings assembled herein comprise a pretty motley crew. (I'm sure Nikki would agree.)

Motliest by far is "Ronsard 58," an early work by Gallic sleaze hero Serge Gainsbourg that I came across on a Top 10 list of Serge's most misogynistic songs. Although unversed en Français, I was able to secure a rough translation of this vaguely jazzy Beat-poet-style blues number. The essence is that Serge is sweet-talking his latest conquest with details of the riches that lie in store for her. Someday, he speak-sings between drags on his Gitanes, this unnamed young woman will have a life of leisure, with cars, boats, and houses. "Mais tu n'seras jamais qu'une petite putain," he assures her, meaning no longer will she be "a little whore." How romantic.

To less salacious songwriters, 58 tends to mean 1958, and not necessarily in a good way. Frickley in South Yorkshire is a small mining town where, according to "Frickley 58" by Chumbawumba, "once the riot coppers beat the pickets to the ground." But a football stadium has since been erected where the protesters were felled, and no one remembers the struggle. Songs like this never become hits. As we all know from the band's lone smash, "Tubthumping," when people in song get knocked down, they get back up again (and declare, "You're never gonna keep me down!"). "Alabama '58" by the Dubliners, another song about injustice, connects the intolerance of the American South to similar ugly chapters from the pages of history.

Fifty-Eight is the name of an African Grey parrot belonging to Hammond B-3 specialist Cochise Jones in Michael Chabon's *Telegraph Ave*, a vinyl lover's dream of a novel that features a cameo appearance by then-candidate Barack Obama. Al Stewart of "Year of the Cat" fame similarly incorporated historical figures into his songs—people like Warren G. Harding, Nostradamus, and Jean-Paul Marat. His "Class of '58" is a cheeky look back at rock's golden age, with the prescient observation that "one day they'll make TV shows on ancient rock-and-rollers." Although not quite ancient, the debut album by Chicago (released under the group's original name, Chicago Transit Authority) featured the eight-minute "Poem 58," which was mainly a showcase for the hot licks of lead guitarist Terry Kath. Although best known for their radio-friendly hits of the '70s and '80s, Chicago sounded far more aggressive on their early albums with Kath. But after he fatally shot himself with a gun he thought was unloaded, Chicago lost its hard-rock edge and headed down the extremely lucrative middle of the road.

If 6 Was 9

THE VERDICT

But you know what they say: you don't need a weatherman to know that if the wind hasn't hit 58 mph, you can't issue a Severe Thunderstorm Warning. And while I can't find a single song in Bob Dylan's canon associated with 58, drawing a straight line from Dylan to Mott the Hoople, the star-crossed champs of #58, is pretty simple. Ian Hunter, Mott's perennially shade-wearing front man, aped Dylan's vocal style for the band's first few records, which made a certain sense given his limited range and the jaded imagery of this former professional songwriter's lyrics. Mott's first few records sold poorly, despite the band's reputation as a titanic live act. Even so, they possessed some of the coolest nomenclature in rock history.

The audacious moniker Mott the Hoople (taken from a 1966 novel by Willard Manus, the term "hoople" meaning "hobo" or "buffoon") was bestowed by their first manager, Guy Stevens, the legendarily mad figure who also named Procol Harum and eventually produced the Clash's *London Calling*. The band also boasted bassist Pete Overend Watts and later added guitarist Ariel Bender (a replacement for Mick Ralphs, who went on to mega-stardom with Bad Company). You just don't see names like that anymore.

Yet the band struggled to translate their onstage energy into recordings. By early '72, with the group at the point of collapse, Overend Watts contacted David Bowie looking for a bass-playing gig. Instead, Bowie offered up "All the Young Dudes" and produced Mott the Hoople's breakthrough LP of the same name.[1] Under Bowie's tutelage, Hunter dropped some of the Dylanesque affectations and picked up a few of Bowie's mannerisms, while the band gained some much-needed studio skills. The Bowie infusion resulted in the resurgence of the band's career as well as one of the great singles of the rock era.

But fame is fleeting, a theme explored in songs like "Ballad of Mott the Hoople." Two years later, internecine squabbling led to the singer's departure, which leads us to "Born Late '58," a song recorded by the remaining members after Hunter left the *Hoople* sessions in a huff.

Not to be confused with "Born in '58" by Iron Maiden lead shrieker Bruce Dickinson, "Born Late '58" is not earthshaking. But it has the signature glam boogie sound of classic Mott and proves Watts capable of singing a lot like his missing bandmate. Eventually Hunter returned to the studio to finish the album. And apparently he approved of the song, in which the singer taunts a would-be suitor who is just a bit long in the tooth to bed the object of his affections: "Admit it, she's greater, shame you weren't born later."

$#@%!

END NOTE

Two minutes and fifty-eight seconds into "Hey Jude," John Lennon can be heard to mutter "fuckin' hell" at his muffed vocal. I kid you not.

A SONG TITLE	AN ALBUM	A BAND	A LYRIC
"58th Street Fingers" – Gap Dream (2012)	*58 Sessions Featuring Stella By Starlight* – Miles Davis (1994)	Article 58 – Scotland, post-punk (early 1980s)	"'58 that was great/ But it's over now and that's all" – Queen, "Modern Times Rock 'n' Roll" (1973)

---

1 Bowie first offered "Suffragette City," but Hunter, who also wanted "Drive-In Saturday," said it wasn't good enough.

59

I firmly believe that a band whose name incorporates the word "anthem" is obliged to render listeners physically unable to keep their fists from clenching and their heads from bobbing to its music. The Gaslight Anthem, punk rockers from New Brunswick, N.J., do right by their name. "The '59 Sound" is the kind of song that would sound great blaring through a car stereo on the Jersey Turnpike beneath a splendidly polluted sunset, or even just ringing through headphones while you wait for your toast to pop. While it may not break any new ground (in fact, there's an unsettling vocal similarity to the Gin Blossoms' "Hey Jealousy"), "The '59 Sound" succeeds. Rock 'n' roll allows for almost infinite variations on a theme, and the force of good crunchy guitars and a sturdy melody will often carry you through with flying colors.

Chickens come in many colors. Imagine, if you will, that a songwriter ended up turning into a chicken. Would you not be tempted to look at his early work for references to laying eggs and clucking? Similarly, when a singer commits suicide, one can't help poring over song lyrics for intimations of his or her self-destructive plans. In that same vein, when a writer comes out of the closet, there is an urge to look back at his body of work for an over-reliance on neutral pronouns. Hüsker Dü provides a case in point; two of the three members of this influential Minneapolis trio eventually came out as gay. (Oddly enough, it was bassist Greg Norton, the one with the handlebar mustache, who was the band's lone heterosexual.) It's hard to say how much of the angst that marked the band's early records was fueled by Bob Mould and Grant Hart feeling forced to live a lie, but the lyrics to "59 Times the Pain" ("Never feeling normal/can't accept the truth/ Resign myself to hating it") certainly speak to the inner turmoil of a deeply conflicted person.

A Swedish hardcore band was so taken with the song that it took the title as a band name and had a pretty successful 10-year run, starting with a single called "Blind Anger & Hate." Well, what did you expect from a band called 59 Times the Pain, "Feelin' Groovy"?

Ah—there's my cue. As with No. 50, a composition by Paul Simon is the proverbial elephant in the room. But unlike the #50 slot, which offered a multitude of choices, this elephant can neither be ignored nor passed over for a more esoteric choice.

THE VERDICT

Simon and Garfunkel's "The 59th Street Bridge Song (Feelin' Groovy)" is the quintessential feel-good song of the '60s, or just about any other era. (Good thing for us Paul used the colloquial rather than the official name for the Queensboro Bridge.) "Feelin' Groovy" is a celebration of what we tend to overlook. It's easy to feel good because you just got paid, just got laid, just met the girl of your dreams, but the source of Simon's groovy feeling is freedom ("no deeds to do, no promises to keep") and the simple sweetness of ordinary things: the morning, cobblestones, flowers, and that lamppost he is moved to address by name. Taking stock,

Simon concludes, "Life, I love you. All is groovy," and somehow it doesn't seem trite or mindless, like the '80s equivalent, "Don't Worry, Be Happy." The lyrical line is so strong that it's easy to forget how much the music swings, courtesy of a rhythm section on loan from the Dave Brubeck Quintet.

Few of the cover versions of the song are half as groovy as the original. In 1967, Harpers Bizarre, then known as the Tikis, was a Santa Cruz surf band led by future Van Halen producer Ted Templeman. The music industry legend Lenny Waronker, who knew catchy when he heard it, approached the Tikis to record the song. Renamed Harpers Bizarre so as not to damage their rep, the band scored a monster hit with a lush, syrupy take, replete with egregiously perky key change. Early in the '70s, a progressive-minded quartet called Cochise decided that "Feelin' Groovy" needed an infusion of heavy, in the tradition of the Vanilla Fudge, who took the wired, sprightly intensity of the Supremes' "You Keep Me Hanging On" and bludgeoned the song into submission. Cochise's "Feelin' Groovy" wasn't quite as misguided, but it shared a common impulse to break a butterfly on a wheel.

The debut LP by Cochise, whose members went on to play with Procol Harum, Foreigner, and Pink Floyd, was notable for cover art designed by future Pink Floyd cover-meister Storm Thorgerson. It depicted the sun rising over the Grand Teton-esque expanse of a woman's naked breasts (quite daring for the time) and probably led to more than a few impulse buys by titillated adolescents. Former street musician Ted Hawkins, who enjoyed a few years of fame before his death in 1995, did a soulful, stripped-down version of the song, and Jimmy Page liked to incorporate the melody into live versions of "Heartbreaker" and "Whole Lotta Love." But the original reigns supreme.

A SONG TITLE	AN ALBUM	A BAND	A LYRIC
"Chateau Lafitte '59 Boogie" – Foghat (1974)	*59 O'Clock* – Dee Jaywalker (2006)	Starflyer 59 – Southern Calif., indie rock (1990s–2000s)	"Fifty-nine cents gets you a good square meal" – Blur, "Magic America" (1994)

GROOVY LITTLE ENDNOTE:

"Feelin' Groovy" may have been preceded by the Mindbenders' "Groovy Kind of Love" and the Troggs' chart-topping "Wild Thing" ("You make everything ... *groovy*"), but Simon & Garfunkel's was the first song to fully exploit the term "groovy," which, along with "far-out," "too much," and "out of sight," vied for essential superlative of the psychedelic '60s. "Somebody Groovy" by the Mamas and the Papas, "Groovy Situation" by Gene Chandler, and many others followed. Years later came the Clash's sublime "Groovy Times" and the acid house gem "Groovy Train" by the Farm, but since then things have slowed down in the world of groovy.

60

The great American composer Aaron Copland said, "If you want to know about the Sixties, play the music of the Beatles." Indeed, the '60s were a decade whose overriding cultural force was a single pop group, yet encapsulating the decade in a single pop song has proven to be a tricky business. T Bone Burnett gave it his best shot in "The Sixties," but despite vocal help from Pete Townshend, the song's social commentary comes off a bit heavy-handed (sample lyric: "Auto dealers don't just sell drive-trains/Sometimes they also deal cocaine").

Sub-T Bone efforts of this ilk include "Sixties Man," a trifle by the Sweet, well past their prime, which scatters allusions to Woodstock and San Francisco over a faux new wave beat; and Barclay James Harvest's "A Tale of Two Sixties," which references Bowie's *Hunky Dory* and *Aladdin Sane* (both from the '70s), serving only to make the Sweet look historically astute in comparison. Perhaps wisely, "Six Six Sixties" by Throbbing Gristle and "Sixties Remake" by Tokyo Police Club employ "60s" as part of an evocative title phrase and leave it at that.

The '60s were good to Nico (formerly Christa Päffgen), whose achievements in that decade included international acclaim as a fashion model, appearing in *La Dolce Vita*, fronting the Velvet Underground on their groundbreaking first record, inspiring Bob Dylan to write "I'll Keep It With Mine," and attending the Monterey Pop Festival on the arm of Brian Jones. She left it all behind in the '70s to forge her own stubborn musical path in collaborations with Brian Eno and others, but her career and personal life took a cruelly downward spiral. Despite producing some critically praised records, she fell into near obscurity and heroin addiction, and spent years living in out-and-out squalor. In 1988 she moved to Ibiza in the hope of turning it all around, but a bicycle accident ended that dream. In her final years, she still sang in the haunted, slightly flat voice that was her signature, adorned only by a mournful harmonium. "Sixty Forty" chills you to the bone like a New York winter: "Will there be another time? Will there be another time?"

Sixty miles an hour is a critical benchmark in the automotive world. Seventies soft-rockers Pablo Cruise worked themselves into an uncharacteristic lather with "Zero to Sixty in Five," which earned a spot in *Guitar Hero II*, but New Order's "60 Miles an Hour" certainly wins the 60-mph crown. Showing off everything this great Manchester outfit does well, it more than lives up to its celebration of the cruising ethos, while suffering slightly from having to follow the godlike "Crystal" both as a single and in the record's song sequence.

In the name of completeness, I offer this set of oddball 60-related songs: "Sixtyten" by Boards of Canada, a spooky yet vaguely funky number from the excellent *Music Has a Right to Children*; "Sixty Sixty," an off-the-cuff instrumental from the late-era Faust; "60%," a spirited blast of pop-punk by NOFX; and "60 Revolutions" from Gogol Bordello, whose lead singer starred in Madonna's directorial debut, *Filth and Wisdom*, and was cheekily described in *The New Yorker* as "explosively hairy." Bob Seger, who is no slouch in the hairy department, gave the numeral a measure of pop immortality in "Night Moves," when he solemnly rasped, "Out in the back seat of my '60 Chevy."

THE VERDICT

Songs about 60 as a rate of speed, a signifier of a decade, or a measure of time are all well and good, but none can match the staying power of "Sixty Minute Man" by Billy Ward & the Dominoes, one of the great sexual boasts in musical history and an unlikely national hit in 1951. It hardly needs to be said that raw carnality was not a staple of the pop charts during the Truman administration. (Chart toppers that same year included Patti Page's "The Tennessee Waltz," "Aba Daba Honeymoon" by Debbie Reynolds, and "On Top of Old Smokey" by the Weavers.) The raw blues has a long history of such references, of course, but when blues songs became crossover hits, the sex tended to be cloaked in metaphor ("I'm like a one-eyed cat peeping in a seafood store") or voodoo-sounding obfuscation ("I got a black cat bone/I got a mojo, too.") Billy Ward and his salacious protagonist, Lovin' Dan—whose sexual prowess and stamina was voiced by the rich bass of Bill Brown, not Clyde McPhatter, the Dominoes' celebrated vocalist—were having none of it. There's not much ambiguity in 15 minutes each of kissing, teasing, squeezing, and blowing one's top.

Many radio stations banned it, but "Sixty Minute Man" went to No. 1 on the R&B charts and became a crossover hit in the guise of "novelty song." But it wouldn't have become a classic if it weren't so downright irresistible, with its sly guitar licks, spirited backing vocals, and delightfully swinging arrangement courtesy of the Julliard-trained prodigy Billy Ward. Sexual boasting is now commonplace, but this recording has a light touch and a sense of playfulness that's never been matched, even by John Lee Hooker, who quadrupled the ante in 2006 with "Four Hours Straight," or the Persuasions, who fessed up to reality with "Can't Do Sixty No More."

A SONG TITLE	AN ALBUM	A BAND	A LYRIC
"60 Minutes of Your Love" – Homer Banks (1967)	*Sixty* – Hugh Masekela (2000)	60 Ft. Dolls – Wales, alt-rock (1990s–2000s)	"Now it's Andy Warhol's time/ Mystic '60s on a dime" – Marianne Faithfull, "Song For Nico" 2002

61

When Roger Maris hit his 61st home run of the 1961 baseball season, breaking the long-standing record held by Babe Ruth, he brought a new measure of acclaim to the 18th prime number. Now that the excesses of baseball's steroid era have effectively obliterated 61's special status to sports fans, the number's association with a certain highway immortalized by Bob Dylan is pretty much set in asphalt. Before we head down that well-trod road, an excursion through some less traveled paths seems in order.

Odense, Denmark, is the birthplace of Hans Christian Anderson, King Canute IV, and the Kissaway Trail, aka the Danish Arcade Fire, an act whose existence lends credence to the music industry truism that if you get big enough, you will be imitated. From the opening banjo lick to the cattle-driver's "*yeahhh*" that sets things off, "61" delivers that ineffable marching-through-the-streets-in-a-ruffled-shirt-playing-kettledrums vibe that Arcade Fire invented. But if you can forget all that and just go with it, the charms are there.

West Virginia is the winter home of Dean Wells, a Vermonter who churns out records under the name Capstan Shafts. Like Robert Pollard of Guided By Voices, Wells holds dear the DIY lo-fi aesthetic, is shockingly prolific, and deals in short songs with absurdist titles (e.g., "Vegans and Meteors," "Bluegene V. Debs," "The Trilateralist Told You Not To"). "61 Sideburns" could pass for the title of a Dali painting, but there's nothing surreal about the song's central line: "We lived in the last genuine time." Wells delivers it with raspy passion, as if he really means it, but in a 2009 email Wells said he was actually taking issue with the line's sentiment. He wrote:

"[It's] the same complaint everyone has as they mature: 'In my day, we had … Music was …' Crap like that. We borrow from the past, use modern conveniences, and still think WE were the real people. I do that anyway."

Wells also shot down my mental image of 31 mutton-chopped dudes, one of them sporting a single 'burn. The title actually refers to an old photo of Mr. Wells that inspired a friend to remark on his beatnik facial hair, "Check out those '61 sideburns."[1]

THE VERDICT

Let's get the rubber on the road. Bob Dylan's "Highway 61 Revisited" ranks up there with his very best. It's hard to imagine that Dylan wasn't familiar with Mississippi Fred McDowell's "Highway 61" or "Highway 61 Blues," but in a 1967 interview with *Rolling Stone* he rejected the idea that there was any major significance to his choice of road. All he would say is, "Highway 61

---

[1] It was refreshing to tease apart the layers of this fascinating miniature and find far less deliberate surrealism than I initially suspected. Then again, an apostrophe would have helped.

exists—that's out in the middle of the country. It runs down to the south, goes up north." Dylan was never a cooperative interviewee, but he knew full well that Highway 61 is a route shrouded in mystery and myth.[2]

"Highway 61 Revisited" is the sound of the newly electrified Dylan at the height of his powers. The song is grounded in the blues, but true to its title, Dylan revisits the old form, and the result is a hallucinatory tapestry wrought from a seedy cast of characters, including Louie the King, the seventh son, and 49 red, white, and blue shoestrings. It all cruises along on a rollicking current of sound augmented with a screaming slide whistle and Mike Bloomfield's siren-like guitar fills, evoking cars and trucks whizzing past on the open road.

Dylan calls it one of his favorites among his hundreds of works, and he has reinterpreted it many times over the years, including a powerhouse version from the legendary "Tour '74" with the Band. Cover versions abound—by the likes of Johnny Winter, the Blasters, Georgia Satellites, and Karen O—all of which have their strong points. Best of all has to be PJ Harvey's bristling assault, an absolute marvel of reinterpretation that clocks in a full 30 seconds shorter but packs even more explosive tension than the original.

A FINAL SHOT OF 61:
Ace Frehley was the 61st guitar player to audition for Kiss in January 1973.

A SONG TITLE	AN ALBUM	A BAND	A LYRIC
"Afterglow 61" – Son Volt (2005)	*Highway 61* – Nash the Slash (1991)	The Sixtyones – U.K., alt-rock (2000s)	"Did you leave it back in '61?" —The Killers, "A Dustland Fairytale" (2008)

---

2 The intersection of 61 and Highway 49 in Clarksdale, Miss., is the crossroads where Robert Johnson was said to have made his deal with the Devil. Elvis Presley grew up near it, and the great blues singer Bessie Smith's fatal car crash occurred on 61. A few years after the Dylan song, Martin Luther King Jr. was felled by an assassin's bullet as he stood on the balcony of the Lorraine Hotel in Memphis, right off 61.

62

"…640**6286**208 821 4808651 32…" —Kate Bush, "π"

It requires real effort to come up with a lyrical instance of 62 in the world of popular song. The fair Ms. Bush doesn't even say "sixty-two," only "six two," but whenever an angelic soprano softly intones the pi sequence up to the 70th digit it's OK in my book. Besides, we're hurting here; there's not even some old nugget, like, say, "Old Blind Joe's East 62nd Street Boogie Woogie," to get us through. Having followed Highway 61 and discovered a deep vein of Americana, taking Highway 62 merely leads us from the U.S.-Mexico border to the U.S.-Canada border, with nary a highway song in between.

Not so for the M62 motorway in northern England, a thoroughfare with a storied past that turns up in songs by the Justified Ancients of Mu Mu and the Human League. "M62 Song" by Doves was recorded under an overpass on the M62. The song is based on the bewitching "Moonchild" by King Crimson, which Greg Lake sang shortly before he became the L in ELP. Fans of Vincent Gallo will recall it from a bewitching scene in *Buffalo 66* that featured a tap-dancing Christina Ricci.

Before we completely abandon the subject of ELP, it seems necessary to address the German trio Triumvirat, known as the German ELP, which is nowhere near as cool as the Beatles of Brazil (Os Mutantes) or even Los Shakers, the Beatles of Uruguay.[1] "The Earthquake 62 A.D.," from Triumvirat's concept album *Pompeii*, is as pompous as its name suggests, with squirrelly keyboard excursions mimicking the exact tone and timbre of the vaunted Keith Emerson. It may even be that it was *this* record—and not ELP itself—that moved Johnny Rotten to upend the bloated '70s rock status quo.

THE VERDICT

Out of this mixed bag, the one to pluck has to be "Rocket Reducer No. 62 (Rama Lama Fa Fa Fa)" by the MC5. Time and again, it has been said that punk rock descended from the Stooges, the New York Dolls, and the MC5. Of the three, the Motor City 5, named for their native Detroit, are the least familiar to today's listeners—and for the record, that includes this writer. Their LPs were the hardest to find (because they went swiftly out of print), and their legacy is the hardest to make sense of.

---

1 The phenomenon extends to individual band members, such as Willem Duyne, aka Mouth of the early-'70s pop duo Mouth & MacNeal, who was known as the Dutch Joe Cocker.

The band's manager was John Sinclair, founder of the White Panther party, which aligned itself with the Black Panther party's goals of empowerment and social equality for black people, while also espousing cultural revolution under the slogan "Dope, guns, and fucking in the streets." At least that's what they claimed to stand for initially. After playing a riotous free show at the 1968 Democratic Convention in Chicago, the group (along with the Stooges) signed with Elektra Records. The following year they released a frenetic concert LP recorded at Detroit's Grande Ballroom, which began with crazy-fro'd lead singer Rob Tyner exhorting the crowd to "kick out the jams, motherfuckers!" At the record company's insistence, the line was changed to "kick out the jams, brothers and sisters" on the record. But the liner notes included the word. As a result, one big store refused to sell it and radio stations wanted nothing to do with it.

Record sales were only part of the problem; the MC5 courted controversy and alienated everyone. The Black Panthers pegged them as "psychedelic clowns." The white radicals considered them insufficiently committed to revolution. Promoter Bill Graham blackballed them after an aborted gig at New York's Fillmore East ended in violence and chaos. And when the record didn't sell, Elektra dropped them.

Which leads us back to "Rocket Reducer No. 62." Every adjective that's ever been thrown at the MC5 applies here: high-energy, revved-up, sweaty, nerve-jangling, incendiary. There is sheer power and precision in the locked-in guitars of Wayne Kramer and Fred "Sonic" Smith, and the rhythm section holds court through the fury with a solid, amphetamine groove. Not to mention Rob Tyner's vocal, which all but screams "Lock up your daughters." It's the confrontational, uncompromising nature of the music that's pure punk: the mania, the volume, the filth and the fury. But the sprawling jamminess of the song, and its pseudo-free-love message, is a far cry from punk's nihilism. Johnny Rotten or Joey Ramone would never have been caught dead singing macho lines like, "I'm a natural man/I'm a born hell raiser and I don't give a damn."

A SONG TITLE	AN ALBUM	A BAND	A LYRIC
"Girl From '62" – Thee Headcoats (1994)	*62* – Paul Anka (1994)	Rule 62 – U.S., pop-rock (1990s)	"Turn that 62 to 125" – Kanye West, "Clique" (2012)

63

Any school kid will tell you that when a donkey and a horse mate, the result is a creature with 63 chromosomes, but thus far songwriters have steered clear of this phenomenon. By a wide margin, #63 songs deal with 1963, sandwiched halfway between rock's breakout year of 1957 and the universally acknowledged death of the '60s at Altamont Speedway in December 1969.

The most inescapable '63-centric song of them all is the horrifically catchy "December 1963 (Oh What a Night)" by Frankie Valli & the Four Seasons, a late-career smash for a man whose string of falsetto-laden hits earned the Four Seasons a place in the rock hall of fame and spawned a musical that seems destined to run for decades. While the song can still cause palpitations among the mom-jeans set, it is suffused with a cloying nostalgia and devoid of any suggestion of the lust that one assumes made the night in question so special. And the piano riff is so jaunty-cheesy it makes Billy Joel sound like Arnold Schoenberg. Teresa Brewer, who ruled the pop charts in the Eisenhower years, was pretty jaunty-cheesy herself. On "Sixty Three Sailors in Grand Central Station," an O Henry-like tale of missed connections, Ms. Brewer goes to the station to surprise her naval beau, but nope. He's somehow slipped away and is headed to her place—to surprise *her*.

Koko Taylor, born Cora Walton and better known as the Queen of the Chicago Blues, has already topped this count-up, nabbing the coveted #29 spot for "Twenty-Nine Ways to My Baby's Door." Thirty-some years hence, she is unbowed in "63-Year-Old Mama"—a defiant boast with an automotive bent—in which Ms. Taylor proclaims she's still got it and never lost it. "The young mens call me a Mercedes, but the old mens say I'm a Jaguar, and their engine don't run cold."

Danny Elfman first honed his film-scoring teeth with *Forbidden Zone*, a slab of oddball cinema circa 1982, starring Herve Villechaize of *Fantasy Island* fame. From its soundtrack comes "Cell 63," a perplexing mash-up of styles that mirrors the film itself, which veered from sci-fi to sex farce and featured lines like, "Flash, be sure and tie your grandfather up and check the knots real good." As a bonus, Elfman makes a cameo appearance in it—as Satan.

THE VERDICT

Once again, this quest has produced an eclectic cavalcade of also-rans and, thankfully, a single choice of unimpeachable aptness. "Rawhide '63" is a retooled version of a 1959 single by the great Link Wray, a half-Shawnee Indian who grew up poor in Dunn, N.C. (not "white-man poor" like Elvis, but "Shawnee-poor," as Wray put it), and who is widely credited with inventing the power chord. As its name implies, this is a blunt, forceful blast of sound created by playing only the low strings of an electric guitar, usually combined with overdrive and distortion. As a child, Wray lived in fear of raids by the KKK, went to work at the age of 10, and learned guitar from a black man named Hambone. Although he came to admire the virtuosity of elegant, clean-toned players like Chet Atkins and Tal Farlow, he lacked the chops to emulate them. So he devised something raw, visceral, and damaged.

Wray's sound is best captured on his signature song, 1958's "Rumble." With fuzzed-out minor chords slow-strummed over a creeping drumbeat that seems to mimic the gait of a hood on his way to a gang fight, it embodies early rock's menace. And Wray—hunched from childhood illnesses, minus one lung from a bout of TB acquired in the Army—performed in a black leather jacket, dark shades, and beaded headband, looking every bit as badass as the sound he made. "Rumble" was an instant smash, even though some radio stations refused to play it simply because of its title and the belief that it might incite impressionable youths to commit real violence. Here's an extremely rare instance of a song being banned simply for its title and sonic implications, as opposed to lyrical transgressions, either real or imagined.

Over the next few years, Wray made numerous singles enlivened by that same overdrive and distortion (initially achieved by poking a pencil through the cone of his amplifier). "Rawhide '63" exemplifies the sound of the late-'50s/early-'60s heyday of the rock instrumental, when a flip of the radio dial might turn up wordless gems in any number of genres, from sweaty Latin-tinged garage rock like the Champs' "Tequila" to surf classics like the Chantays' "Pipeline" to the soulful swing of "Green Onions" by Booker T & the MGs. Built around simple blues chord changes, "Rawhide '63" features the surf-style drumming of Wray's brother Vernon, a pumping keyboard, and Wray's inimitable licks, combining touches of Chuck Berry and Duane Eddy but adding up to a sound all his own.

Every notable guitarist in rock has traded in the feedback, noise, and distortion that were Wray's essential palette. There is no style of rock played in the past 50 years that is not indebted to the singular power of a blasted-out guitar chord, which began with Fred Lincoln Wray Jr.

A SONG TITLE	AN ALBUM	A BAND	A LYRIC
"63" – Theatre of Hate (1996)	*A Scant Sixty-Three* – The Great Outdoors (2005)	The 63 Crayons – Virginia, pop-rock (2000s)	"I don't suppose you would remember me/ But I used to follow you back in '63" – The Who, "Bell Boy" (1973)

LAST WORDS:
"If I could return in time and see one band live, it would be Link Wray and the Ray Men." —Neil Young

"The most devilish thing is 8 times 8"
—Marjory Fleming (1803–1811)

Marjory Fleming found the product of 8 and 8 extremely vexing, and she might have been on to something. This child poet, writer, and diarist from Kirkaldy, Fife, Scotland, was a favorite of Sir Walter Scott, and she showed prodigious literary skill in her brief life. There *is* something devilish about 64, at least where the Beatles are concerned. You might say that "When I'm Sixty-Four" is to the Fab Four what *Macbeth* is to Shakespeare.

I make this allegation because several of the best-known warblers of "When I'm Sixty-Four" never got close to the age in question. They include Keith Moon, who sang it on the misguided soundtrack to the Beatles hodgepodge *All This and World War II*; John Denver, who died in a plane crash at age 53; and former child actor Jack Wild, who co-starred as a plucky castaway in *H.R. Pufnstuf*. Others who covered it experienced high-level personal tragedy: the pop singer Claudine Longet was convicted in 1976 of manslaughter in the death of her boyfriend, skier "Spider" Sabich, and never performed again. British singer and performer Georgie Fame sang the song, and years later his wife, the former Nicolette Harrison, Marchioness of Londonderry, leaped to her death from the Clifton Suspension Bridge in Bristol. Parodying the song, however, does not seem to have any associated risks. The Rutles did a spot-on pastiche called "Back in 64," and to my knowledge, the lesser talents who came up with "When I'm 84" and "When I'm 43" have not incurred the heavy hand of fate.

THE VERDICT

"When I'm Sixty-Four" is a modern standard, instantly recognizable by young and old, yet it's one of the most atypical songs in the Beatles canon. Not because it doesn't rock—from the very start, the Beatles showed a willingness to make forays into a number of non-rock styles: the gentle balladry of "And I Love Her," the squeaky-clean "Till There Was You" from Broadway's *The Music Man*, and onward to "Yesterday," which convinced many skeptics (or squares) that the Beatlemania was more than a passing craze. What makes "When I'm Sixty-Four" singular is that it *sounds* old. Yes, they would repeat this trick with "Your Mother Should Know" and "Honey Pie," but on the screamingly psychedelic *Sgt. Pepper*, the song stood out, in the words of author/Beatles savant Ian MacDonald, "like a comic brass fob-watch suspended from a floral waistcoat." Indeed, sandwiched between Harrison's "Within

You and Without You" and the shimmering colors of "Lovely Rita," it constitutes a hard 180-degree turn into the musty, more innocent past.

"When I'm Sixty-Four" was almost wholly a McCartney creation. John contributes some acoustic guitar in the last verse; Ringo provides minimal drums and some chimes; and Lennon and Harrison sing background vocals (their "ah-ah-ahs" after "you'll be older too," as Tim Riley points out in *Tell Me Why*, are the aural equivalent of a grown-up's finger wag). But Paul composed and sang it, played the piano on it, and wrote it for his dad, Jim. The song had kicked around since the band's days in Hamburg, where they would play it as an instrumental during technical difficulties. The pronounced oompah-band vibe must have added interesting visuals for young people using *Pepper* as the background for an LSD trip. Yet it's hard to imagine that it was any kid's favorite track on the album, and just as hard to imagine it *not* being the favorite song of those in Jim McCartney's generation.

Few writers have attempted a #64 song of their own. The most interesting item I've come up with is an obscurity lover's dream: a song by *the former band* of a one-hit wonder. It's not quite as juicy as, say, something by the teenage garage band of the guy who did "They're Coming to Take Me Away Ha Ha," but it's close. Roger Jouret, better known as Plastic Bertrand, had a European smash with "Ca Plane Pour Moi," a delirious bit of doggerel built around Jouret's nonsensical French-and-English rant and its wordless, pseudo-Beach Boys hook. I tell you this because previous to his brief moment as a hitmaker, Jouret played drums for Hubble Bubble, which produced two undistinguished records in the late '70s. "Number 64," from *Faking*, is an innocuous folk-rock/new wave hybrid, with ringing acoustic guitar chords and a typical chugging eighth-note groove.

A SONG TITLE	AN ALBUM	A BAND	A LYRIC
"Spirit of '64" – Bocky & the Visions (1964)	'64–'95 – Lemon Jelly (2005)	64 Spoons – Watford, England, prog (1976–1980)	"Hangin' out, rollin in my '64" – Dr. Dre, "Rat-Tat-Tat-Tat" (1992)

65

As the U.S. speed limit and the age at which you join the ranks of the elderly, 65 comes off as a scold, a cut-off point—certainly nothing to celebrate. Yet it has its advantages. Country great Merle Haggard addressed this on "Come On, Sixty-Five," a musical wish to hasten the arrival of his 65th birthday so he can get his gold watch, kick back, and perhaps enjoy some warm evenings sipping Bourbon and branch on his porch. The song is summed up in the line, "I've heard it said that hard work never did a body's body any harm/Well they were wrong." In "Southern Pacific," Neil Young's aging train engineer agrees: "When I turned sixty-five/I couldn't see right." A further echo of this lonesome-train feeling can be heard on "Sixty-Five Days," a reverb-y instrumental by the Knoxville Girls, who named themselves after a chilling murder ballad popularized by the Louvin Brothers.

The doo-wop era began in the early '50s and was buried under an avalanche of Beatlemania some ten years later, which is why Paul Davis's egregiously sunny hit "'65 Love Affair" is such a historical travesty. This staple of rock radio circa 1982 (and a favorite of Dick Clark) is a slice of pure heaven for anyone who considers "doo-wop-diddy" the most joyous sound in creation. But, by actual 1965, when the Who sang "My Generation" and the Byrds recorded "Eight Miles High," this kind of ditty was conspicuously gathering dust.[1] Originally titled "'55 Love Affair," Davis's song got a name change when some clever A&R people decided that '55 sounded dated—not that the radio-listening youth of America were sticklers for accuracy. The spin-doctors got it right, though; it ascended to the Top 10, and the Jesus of Nazareth look-alike had a giant hit on his hands.

White Zombie's commercial breakthrough, "Thunder Kiss '65," would make a fine anthem for an army of marauding Huns, with Rob Zombie's brawny vocals leading the way over turgid, bottom-heavy riffing. In the midst of the maelstrom, which incorporates film samples (*"I never TRY anything; I just do it. Wanna try me?"*), police sirens, and shredding axe work, it's easy to miss the similarity to Led Zeppelin's "Immigrant Song." It's also easy to miss what it has to do with 1965, but a lyric sheet proves that *What's New Pussycat* and "Satisfaction"—both produced in 1965—are name-checked. But this is not a song for study; it's meant for head banging or pole dancing, and on that front it is highly recommended.

Gene Chandler never topped "Duke of Earl," a vocal tour de force and an undeniably great single of the early rock era. But the Chicago native, born Eugene Dixon, had a subsequent string of Top 40 hits produced by the great Curtis Mayfield, including "Rainbow," aka "Rainbow '65." Consisting of Chandler's ad-libbed vocals over a trilling piano and measured drumbeat, the song served as a trusty encore at live shows. In the version recorded at the long-gone Regal Theater in Chicago, titled "Rainbow '65 (Parts 1 and 2)," the crowd keeps responding, including an emphatic exhortation of, "You *go*, dad!" When Chandler confesses, "*I gonna reach out and-uh BITE-cha,*" delirium erupts.

---

[1] True, Manfred Mann hit No. 1 with "Do Wah Diddy Diddy" a year earlier, but its spirit was far more early-'60s pop than '50s doo-wop.

THE VERDICT

Darling Downs, an Australian duo, comprises two mainstays of Sydney's indie rock scene. Ron Peno, the former lead singer of Died Pretty, and Kim Salmon, who's credited with forming one of the first punk bands in Australia, the Scientists, have spent most of their lives playing music influenced by the Stooges and the Velvet Underground. But together they make a sound rooted in Americana, an acoustic mixture carved from Salmon's banjo and guitar, Peno's rich and resonant vocals, some harmonica and the occasional shake of a tambourine. "Circa '65" resulted from the duo's idiosyncratic method of songwriting. As Salmon explained to me in an email, he starts by laying down an instrumental groove, and Peno free-associates over the top in search of a melody.

"The stuff he sings is random and tends to borrow from rock's rich tapestry," wrote Salmon. "One of the lines for this song ended with the phrase 'back in 1965,' because, as your essays testify, there is a history of numbers, particularly dates, in rock lyrics. For example, Jonathan Richman's 'She Cracked' has the lines, 'Well she was sensitive /She understood me/ She understood the European things of 1943.' This is definitely the type of feel that Ron was looking for."

Salmon says the only downside to his partner's preference for improvising words, rather than writing them down, is that it comes down to Salmon to decipher and transcribe Peno's spontaneous guide vocal.

"He's lazy, and he thinks this is easier," said Salmon. "Neither Ron nor I knew exactly what had been sung, so translating it in itself made the end results even more surreal, e.g., 'I had to have the halo, when I hit the floor/I've been on a timeless journey since 1964.' It was like Ron was on the couch and I was his analyst, and then William Burroughs got the results and cut it up!"

And in the spirit of William Burroughs, the randomness of the process produces something with its own unique resonance. "I have found working with Ron that even though what he does is somewhat random, it does tend to take on a huge amount of meaning simply because he doesn't allow himself the chance to contrive. Everything he sings comes straight out of his psyche. By its sheer meaninglessness, 1965 has become full of mystery and spiritual significance."

A SONG TITLE	AN ALBUM	A BAND	A LYRIC
"65" – Josh Rouse and Kurt Wagner (1999)	*The Sound of 65* – The Graham Bond Organization (1965)	Q65 – Holland, garage beat (1960s)	"You could be laughing 65 percent more of the time" – John Grant, "GMF" (2013)

66

Like the Beatles' "When I'm Sixty-Four," which seems to have scared off sensible songwriters from writing another #64 song, the winner in this category is a colossus that dominates the competition all but completely.

In 1946, Bobby Troup was akin to a journeyman pitcher in baseball, someone with the goods to make it to the majors but lacking the X factor to ascend to the level of the greats or near-greats. Nevertheless, Troup was a man of many talents. An able pianist and a recording artist in his own right, he was also a record producer, TV show host, and an actor. (He portrayed bandleader Tommy Dorsey in *The Gene Krupa Story*, among other movie roles.) But Troup seems to have had a sense of his own limitations; though he wasn't quite leading-man material, it didn't stop him from acting. He found steady employment on shows like *Mannix* and *Dragnet* and, most notably, had a featured role as Dr. Joe Early on *Emergency*, where he worked alongside his second wife, the blonde-tressed torch singer Julie London, who played hot nurse Dixie McCall. But let's face it: By the time he was well ensconced on the tube in the early '70s, Troup could have retired on the royalties from a song he wrote right after World War II. In a radio interview from 1985, Troup described the song's genesis:

"My wife [Cynthia Hare] and I were eating in a Howard Johnson's and looking at a road map … She said, 'Why don't you write about Route 40.' I said, 'That's silly, because we're going to pick up Route 66 outside of Chicago and take it all the way to Los Angeles.' She said, 'Get your kicks on Route 66.' I said, 'God, that's a marvelous idea for a song.'"

Troup finished it in the car.[1] When they arrived in L.A., he played the song for Nat King Cole, who seized on it immediately, and his version went to the upper reaches of both the R&B and pop charts. It was by far Troup's greatest contribution to American culture—but he was no one-hit wonder. He also penned "The Girl Can't Help It," a Little Richard screamer that served as the title to a seminal early rock 'n' roll movie, as well as "Their Hearts Were Full of Spring," which the Beach Boys recorded, and "The Meaning of the Blues," recorded by Miles Davis during his golden age.

Some history: The route in question was first laid out as a wagon trail, with a delegation of camels in tow, in 1857. Designated no. 66 in 1926, it became a key route for the westward migration of Dust Bowl refugees, a process chronicled in *The Grapes of Wrath* by John Steinbeck, who dubbed it "the Mother Road." Over the next few decades, Route 66 became a critical cross-country thoroughfare in an America still in the throes of its love affair with the automobile, as well as a breeding ground for the development of the modern filling station. Perhaps inevitably, though, Route 66 was not cut out for America's postwar prosperity; the four-lane interstates were better equipped to handle heavy-duty trucking, and the road swiftly deteriorated physically as it shrank in importance. By the '70s, major stretches shut down, and in 1985 it was officially decommissioned as

---

1 His marriage to Cynthia didn't last, but he was gracious enough to give credit where it was due.

a U.S. highway. Today there is a movement afoot to preserve parts of the road for its cultural importance. Bobby Troup's song helped cement Route 66's status as an American icon in the public consciousness.

"Route 66" is an amazingly versatile song; it works in just about any genre, from bossa nova to a primitive electric stomp. Hundreds of people, from Ray Charles to Anita Bryant, have been moved to cover it. Troup's jazzy original showed off his keyboard chops and sly scat singing. Nat Cole ran with it, adding his mellifluous phrasing and rich rasp to Troup's gorgeous syllables and kicked the thing into the stratosphere. Numerous versions followed: big band style (Harry James, Bing Crosby) and lighter takes in the Cole vein (Mel Tormé, Louis Prima, Louis Jordan).

Chuck Berry's 1961 version is perhaps the earliest straight-up rock take on the song. Given his deep influence on the Rolling Stones, it would make sense to surmise that it was Chuck who inspired their cover. But, according to several accounts, it was actually the rendition done by soporific crooner Perry Como that the lads studied. Nevertheless, the Stones transformed this slinky concoction into a fierce, groovy rocker on the strength of Keith Richards' Berry-esque rhythm and lead lines; the tight, brisk rhythm section, urged on by handclaps; and Mick Jagger's brash vocal. (He stumbles a bit on "don't forget Winona" and obviously couldn't care less.) The start-stops in the bridge amp up the tension and release, while the chunky guitar lick that anchors the song has been incorporated into practically every subsequent cover, from garage/pub rock offerings by the Count Bishops and the Eyes to faithfully Stones-y versions by the Pretty Things, Tom Petty & the Heartbreakers, the Replacements, and R.E.M. The Cramps' hushed deconstruction is an exception, but even covers by goth proponents like Depeche Mode and Lords of the New Church owe a great deal to the Stones. An old-school honky-tonk version by Asleep at the Wheel, Buckwheat Zydeco's N'awlins-flavored rendition, and the U.K. Subs' hardcore bash-o-rama demonstrate the infinite variations the song can withstand. Amazingly, given the visceral, trip-off-the-tongue nature of the lyrics, which border on poetry, no one has seen fit to do a rap version.[2]

THE VERDICT

While Nat King Cole's is the definitive cover of the original, the Stones turned "Route 66" into a lean, mean slice of visceral rock 'n' roll. Thus, with all due respect to the sophistication and subtlety of Mr. Cole, my deep-seated propensity to rock compels me to confer top-ranking upon the Stones' cover of Bobby Troup's gem.

A SONG TITLE	AN ALBUM	A BAND	A LYRIC
"66" – Afghan Whigs (1998)	*Sixty-Six to Timbuktu* – Robert Plant (2003)	Salem 66 – Boston, indie rock (1980s)	"Down Route 66 you pilfered love and green stamps" – Magnetic Fields, "The Desperate Things You Made Me Do" (1995)

---

2 Although Public Enemy did touch on 66-ness in "Incident at 66.6 FM," a brief collage made up of racist comments to a radio call-in show.

67

Your local mathematician will tell you that 67 is what's known as a lazy caterer's number, meaning it's part of the so-called lazy caterer sequence. This phenomenon refers to the number of pieces of a round object—say, a pizza—that can be made with a specific number of straight cuts. For example, if you make three straight cuts that meet in the middle of the pie, you get 6 slices, but you can make 7 if they don't meet in the middle. Thus, a lazy caterer who knows what he's doing can make 11 *strategic* cuts in an enormous pizza (imagine he's catering a wedding) and make 67 slices. This particular lazy caterer used a tri-stroke and yielded the following seven slices:

- "Route 67" by Let's Active is related to Route 66, but not the one you're thinking of. Mitch Easter wrote to me, "One of my favorite songs is the *Route 66* TV theme—the Nelson Riddle composition, not the 'get your kicks' song. But you know how in the U.S.A. there's always this romanticism for 'the road' and this kind of blues vibe, which invariably gets associated with a limited number of states. Since the track 'Route 67' seemed to have a certain Blues, and therefore Road factor, I guess, I thought I'd give N.C. a nod as having some viable Roads, too. Mainly it was just a mild (very mild) joke."

- Elton John's "Old '67" is a hokey return to a "time of innocence" from the unnecessary sequel to *Captain Fantastic and the Brown Dirt Cowboy*.

- "Questions 67 and 68," the first single released by Chicago Transit Authority, derives its title from the love life of keyboardist Robert Lamm, whose girlfriend besieged him with commitment inquiries during the last two years of the Johnson administration.[1]

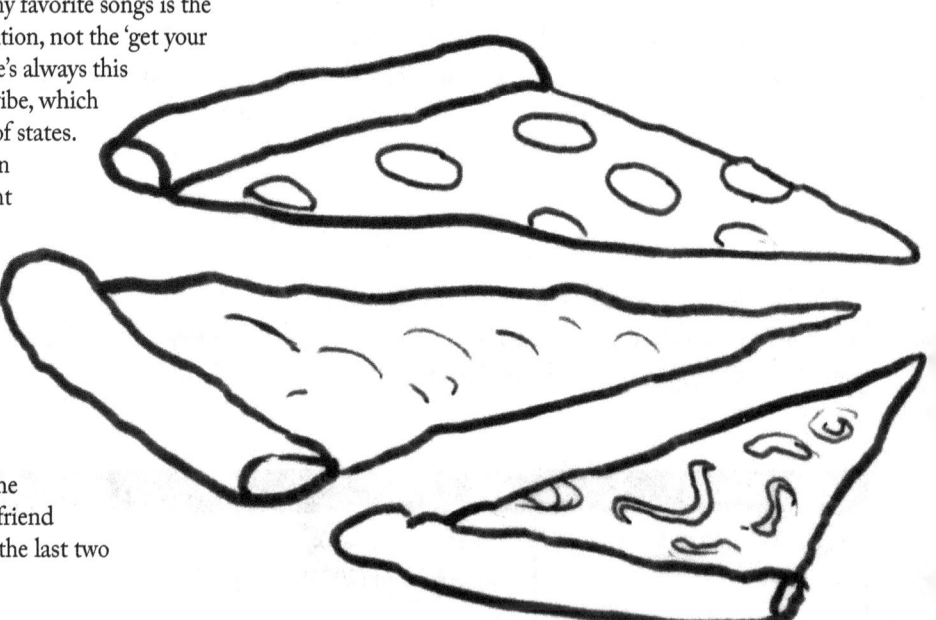

---

[1] Say what you will about Chicago, the group has sold more records than any American band but the Beach Boys. And while many of the hits were formulaic Me Decade radio pabulum, Chicago initially forged a sound from the disparate influences of jazz, classical, and rock. The band also deserves credit for its mathematical bent; in addition to a pair of juicy numerical singles (the other one being "25 or 6 to 4"), Chicago numbered its album titles sequentially.

If 6 Was 9

I really *wanted* Easter's mild (very mild) joke to be my #67. It's a hot blues-country jam, no doubt, and the song's source, *Big Plans For Everybody*, meant a lot to me during the summer of 1986, when I waited tables at a sushi restaurant in the East Village. But when I listened to mixtapes on my Walkman while cleaning the windows before a shift, it was the triumphant opener, "In Little Ways," that I'd be cranking, or the gentle acoustic dreamscapes of "Badger" or "Reflecting Pool" when I was swooning over some un-have-able love object. I never rated "Route 67" as highly. As the record's lone instrumental and final song, it always seemed at odds with those moody, stained glass folk songs or the psychedelic stompers that earned praise from Robert Plant at the time. I liked it better in 2009 than in 1986, but I couldn't advocate lustily for it as the ultimate #67. And that kind of sucked, because obviously Elton, Chicago, and Driver 67 were out. So that left me with the Sam Roberts Band, the Infadels and Love Battery. Good songs all, but something was missing, namely my deep enthusiasm.

Back I went to the well, where I discovered that "67," by the Korean pop duo As One, was not the answer. I did discover "67," by Lord Kitchener, a Trinidadian calypsonian whose first musical success was a recording of the traditional "Mary, I Am Tired and Disgusted," under the title "Green Fig." But I was truly stuck. Then, in the midst of these depths it struck me—like a pizza thrown by a caterer suddenly shaking off his ennui—square in the face: I live on the bloody Lower East Side of New York. A working

- "Car 67" by Driver 67, a U.K. novelty hit in 1978, dares to posit David Bowie's "Space Oddity" not as an intergalactic conversation between ground control and Major Tom but as colloquy between a beleaguered London taxi dispatcher and a lovesick cabbie, set to a spindly faux reggae beat.

- Love Battery's "67" gleefully kicks '60s nostalgia to the curb with Stooges riffs and a sneered vocal by singer Ron Nine.

- "Detroit '67" finds the Quebec-born Sam Roberts and his band working a well-worn honky-tonk piano groove as Roberts yearns for a Motor City he's too young to remember.

- "Jagger '67," the Infadels' twitchy, electropop tribute to a girl with "twisted hair" and "lip-ring stare," gets its kinetic drum kick from Alex Bruford, son of prog giant Bill Bruford.

musician tends bar just a block away. If the world cannot supply me with a suitable #67 song, I would have to have one made. Zee, a shaven-headed New Zealander born Zeebling Monroe, who worked at Barramundi on Clinton Street, was up to the task. He had played me some of his sample-heavy electronica stuff at the bar, and it was pretty damn good. I rubbed my palms together: This *just might work…*

Long story short: Zee agreed to do it, and I moved on to Chapter 68. A few months later Zee sent me an MP3 file for "Hong Kong '67." The song has an ominous atmosphere and was built around sound bites from a pair of Wong Kar-Wai films, *In the Mood for Love* and its sequel, *2046*. Zee even directed a video for it.

For a good five years, Chapter 67 ended with Zee. I genuinely liked "Hong Kong 67" and the crazy way it had come about. Sure, I had kind of pulled a fast one by commissioning a song and crowning it king, but I figured it was an anomaly, a desperate measure borne of a desperate time. And hell, what were the odds that someone would write a new #67 song that would force me to revisit this chapter anyway? Then, in summer 2014, there it was, above the skies of London: a chartreuse blimp emblazoned with the logo of Aphex Twin.

THE VERDICT

The press release-as-dirigible heralded a new 12-song collection called *Syro*. The near-mythical producer and electronica pioneer born Richard D. James had not released any new music in 13 years, and his return to the scene was met with the sort of rapture that would have greeted Picasso after a lengthy hiatus from painting.

The first song to waft out from behind the curtain was "minipops 67 [120.2][Source Field Mix]."[1] It sent a charge through me. Even before hearing a note, I felt positive that I had found my #67. No way was Aphex Twin going to unleash anything less than killer for the lead track from his comeback LP. Yes, I had some misgivings in having to bump Zee's song, which I'd earnestly solicited and had been fully prepared to stick with, despite its undeniable touch of nepotism. But aesthetically speaking, selecting "minipops 67" is pretty unimpeachable. Any sane person who works in something called electronica (Zee, for example) would defer to the master. There's also a bit of mojo at work here. In numerology, 67 becomes the sum of 6 + 7, and 13 is equal to the number of years that James took off before returning with *Syro*.

During his exile, James retreated to rural Scotland, got married, and had kids; the home and hearth vibe shows in the music. *Syro* is still a thing of Pynchonian complexity, with its share of unsettling textures, but the caustic edge has ebbed, leaving the music's otherworldly beauty to shine that much brighter.

James has always played with semiotics and signifiers. His early compositions were named for complex geometric shapes or made-up terms that resembled ancient Greek computer code ("Heliosphan"). He worked anagrams of Aphex Twin ("Next Heap With," "The Waxen Pith"), used chemical and pharmaceutical names like "Isopropanol" and "Ventolin," and on *Drukqs* the titles were unpronounceable ("Jynweythek," "Bbydhyonchord"). Recording under various names (Richard D. James, Aphex Twin, AFX, Caustic Window) created a further impediment to fathoming the person behind the sounds. With every song on *Syro* named for

---

1   The beat count of each song on *Syro* appears in square brackets within the title, e.g., "produk 29 [101]."

the vintage gear that James uses on it, the record is a departure from past inscrutabilities.

On "minipops 67," James relied on the Korg Mini Pops, an early, no-frills rhythm machine with a limited number of preset rhythms, whose most prominent usage pre-2014 was by French composer Jean Michel Jarre in his classic of '70s synth wizardry, "Oxygene." Somehow, like Segovia picking up a dime-store guitar, James bends the Mini Pops to his will and renders a luxuriously rich, ultra-modern sounding track. Manipulated voices intone like cyborgs trying to croon, heavy-liquid keyboards glint in matte finish, and a bass line bounces a steady, danceable groove. Yet rather than sounding like a math problem solved on a sequencer, the track has a sense of play to it, if not outright playfulness. James himself calls *Syro* his pop record, "or as poppy as it's going to get."

A SONG TITLE	AN ALBUM	A BAND	A LYRIC
"67" – King's X (1996)	'67 – Lord Kitchener (1967)	67 – London, punk pop (1990s)	"67, 67/ Freaks dance like they in heaven" – Prince, "Now" (1995)

'68

Pink Floyd's ascendancy to Monolithic Rock Act was nothing like a done deal in the years after founder Syd Barrett wandered off into the acid mist. With David Gilmour picking up guitar duties, the band continued with the long-form compositions that would mark their later work. But before the breakthrough of *Dark Side of the Moon*, in 1973, Floyd had yet to hit upon a sound that could claim the masses. "Summer '68," written by keyboardist Richard Wright, appeared on 1970's *Atom Heart Mother*. Wright, who would later be consigned to the band's margins, was at this point still allowed to sing lead vocals. As such, the song captures the band in transition.

"Summer '68" is Pink Floyd's kiss-off to the excesses of the late '60s, a rejection of the scene and the false intimacy of free love.[1] It starts in familiar Floydian sighing-pastoral mode, with just Wright's keyboards and hushed vocals. The singer speaks directly to the girl he's just spent the night with, in decidedly unsentimental fashion: "Would you like to say something before you leave?" Soon things shift, rather jarringly, like the song's been dosed, to a swirling section with Beach Boys-like interlocking vocals intoning, "How do you feel? How do you feel?" In comes a bombastic, nautical-sounding brass theme, only instead of suggesting a graceful sloop à la "John B," this feels like a foundering ship taking on water under a roiling sky. The song alternates between these sections, culminating with the declaration, "I've had enough for one day."

The Welsh foursome known as the Alarm never really broke through in the U.S., but in Europe the band had an impressive number of hits and an army of loyal fans. Critics called them a second-tier U2. Joe Strummer called them "the imitation of a shadow of the Clash." Even rabid fans of the band (severe Alarmists?) would acknowledge that every Alarm song is a call to arms. Such a strategy may be limiting, but "Sixty-Eight Guns" does what an anthem sets out to do. Sure, they liberated a few Spaghetti Western-style brass touches from *London Calling*, and the line "Up on the terrace I can hear the crowd roar" is shameless, audience-baiting fodder for live shows, but it all works here, and Mike Peters, landing somewhere between Billy Idol and Bono, belts out a rousing vocal.

THE VERDICT

The "Sixty-Eight Guns" vs. "Summer '68" matchup is similar in spirit to the #51 contest, wherein a great song by a really good band (New Model Army) vied with a lesser track by one of the greats (Jimi Hendrix). In that spirit, I lean toward the Alarm. It's

---

1 A similarly titled song by the Charlie Daniels Band, "Summer of '68," takes a dim view of the countercultural shifts of the era as well as liberal politics in general ("Save the whales and kill the babies"). And to think that Daniels' first national hit, "Uneasy Rider," was a sympathetic story song about a longhaired, pot-smoking hippie who gets hassled by some good ol' boys in a Mississippi watering hole.

way more 68ish; the number is right there in the chorus, while it goes unmentioned in the Floyd. And the sturdy Springsteen-ian chord progression and gang-vocaled "yeah-oh's" add up to a song that, despite its obvious overwroughtness, always has me ready to take up arms by the final chorus.

A SONG TITLE	AN ALBUM	A BAND	A LYRIC
"68" – Jawbox (1994)	'68 – Santana (2003)	The 6ixty8ights – U.S., indie (2000s)	"And the sixty-eight summer festival wallflowers are thinning" – Elton John, "Captain Fantastic and the Brown Dirt Cowboy" (1975)

69

Stephin Merritt of Magnetic Fields named his epic work 69 Love Songs because he needed an outsize numeral. After considering 100, he settled on 69, both for its graphic potential and because it was the next lowest number below 100 "that had another relevant meaning." And he's right. Sure, 72 is the number of beats per minute in the average human heart rate, and 98.6 is the average human body temperature, but 69, nimble shorthand for an iconic—not to mention visually palindromic—act of mutual oral gratification, is the indisputable big kahuna between 50 and 100.

It's funny, then, that of the scads of 69-titled songs out there, those that specifically reference the act itself are relatively rare. At the forefront is T-Pain's "69." "I've been doing tongue exercises," he promises, before making the anatomically impossible claim, "I spread that booty so wide/I can tell that shit's spread by the look in her eyes," next to which Rick James's "She Blew My Mind (69 Times)" is a model of restraint. By and large, though, the trend continues: Most 69-titled songs refer to a year—in this case the year of Woodstock and the moon landing, the Manson killings, and the Miracle Mets.

The grizzled French bard Serge Gainsbourg, who started life as Lucien Ginsberg, saluted the end of the '60s with the languid "69 Année Érotique." A duet with his lover Jane Birkin, this collaboration is tamer than the record's controversial hit, the softcore moan-fest "Je t'aime…moi non plus,"[1] but the lusty message is never in doubt. Unlike with "Je t'aime," the Vatican did not denounce "69 Année Érotique," presumably because the double entendre of the song's "*soixante-neuf*" chorus hook was too subtle. Mick Harvey of the Bad Seeds paid homage to Serge in the mid-'90s with *Intoxicated Man* and *Pink Elephants*, two collections of Gainsbourg songs sung in English, including, naturally, "69 Erotic Year." (Eet sounds better en Français, non?)

If you are hard-pressed to come up with a Polish rock band, you're not alone. The Polish electronica/dance music scene has made inroads stateside, with Kraków's long-running Unsound Festival reaching U.S. shores in 2010. Still, most of Poland's music scene is largely unknown to most Americans, myself included. That's why I was so intrigued to discover "Sixty-Nine Moles" by George Dorn Screams, a Warsaw-based quartet that describes itself as "Joy Division meets Mazzy Star," which isn't a bad description. I've listened to the song at least 20 times, and I still don't know what the 69 moles signify. Nor has my research revealed the identity of George Dorn, but no matter: a little mystery goes down just fine with a sound this dreamy.

Star 69, aka "call return," has been a telephonic option since the early '90s. Besides unleashing an untold tsunami of awkward moments, the catchy shortcut also inspired a number of songs and at least one band name. R.E.M. was there early with "Star 69" from *Monster*, that clamorous blast of glam and guitar noise that would be the band's last recorded work to approach sales expectations. "Star 69," a song about persecution via telephone, barrels forward with added propulsion from Michael Stipe's overlapping vocals, which bring a touch of youthful chaos to the proceedings, while in Fatboy Slim's thumping house track of the

---

[1] Some believed Ms. Birkin's orgasmic trilling was authentic, making "Je t'aime" the "Kiss Kiss Kiss" of its day. The controversial song was a major to-do in Europe, but in the U.S. it stalled on the charts, appropriately enough, at #69.

same name, the entire lyric consists of a vocal sample from the Roland Clark song "I Get Deep" ("They know what is what, but they don't know what is what, they just strut. What the fuck.") There's no mention of phones, but the lyric is repeated at least 69 times.

Sixty-nine seems to stoke the fires of lust and desire, and sometimes it unleashes the powers of hell. Ministry's "Psalm 69" is a spiky tapestry of creepy film samples, mock-sermon sound bites, pummeling guitar riffage, and Cookie Monster vocals (before they became a staple of various schools of metal.) Fusing the most aggressive elements of hard rock, techno, hardcore, and industrial, this is one ingeniously well-calibrated death machine of a song.

To the man on the street, there is an obvious answer for #69 primacy: "Summer of 69" by Bryan Adams. For sheer name recognition, it trumps every song on this list. Now, I'm not going to fault it for being erroneously autobiographical. Why should it matter if Adams would have been about 5 years old in 1969 when, according to the song, he's making some pretty grown-up vows on his baby's mama's porch? Not an issue; let's just say he's playing a character. Besides, Bryan Adams is no stranger to controversy—he seems to divide people. To some, he's utterly derivative, a third-rater. Others find his rough-hewn pipes and meat-and-potatoes rock really hit the spot. But the song just never sunk its hooks into me. Adams said its inspiration was Bob Seger's "Night Moves," and while the two share the young-fumblings imagery, Adams sounds packaged and pat compared with Seger's classic. (On the other hand, "Run to You" and "Cuts Like a Knife" still have the power to impress this foe of corporate rocking.)

THE VERDICT

The biggest mystery behind the winning song for this hotly contested spot is why Captain Soul's debut single, "T-Shirt 69," is about as well known here as the Polish indie rock scene. True, the band (named after a Byrds instrumental) missed power pop's brief mid-'70s heyday by about 25 years, but "T-Shirt 69" is so downright stunning it's an injustice the song never even found its way into, say, some lame movie with Liv Tyler. While the song's origin is fairly banal—a girl wearing a 69-emblazoned T-shirt caught singer Adam Howorth's eye—the captivating result of that fleeting encounter is a feat of musical alchemy: turning a sad, empty feeling into four minutes of sheer glory.

Captain Soul was an English foursome that spent three years in contractual limbo with Sire Records before joining the Poptones label, established by Creation Records founder Alan McGee, in 2000. The band's name attests to one of its major influences, but as with Teenage Fanclub, the band's sound conflates the Byrds with Neil Young's whomping guitar overdrive and the wounded heart of Big Star. And it's the spirit of romantic yearning, which courses through so many of pop's great singles, that gives "T-Shirt" its urgency and poignancy: "I'm at your feet," croons Howorth, "but you're out of reach."

A SONG TITLE	AN ALBUM	A BAND	A LYRIC
"Butter of '69" – Butter 08 (1996)	69 – A.R. Kane (1988)	Sham 69 – England, punk (1970s)	"I'm-a need 69 real bad girls for my tour bus" – Nicky Minaj, "Girls Fall Like Dominoes" (2010)

Boards of Canada, a Scottish duo that knows its way around the esoteric, can lay claim to writing the world's only ode to the number 70: "The Smallest Weird Number." This brief, floaty instrumental takes its title from 70's dubious distinction: In mathematical parlance, a weird number is "abundant" but not "semiperfect," and 70 happens to be the smallest weird number.[1] Thus, as I see it, 70 is like a person with all the right connections (to the other decade numbers, to his family of abundant numbers) but, lacking semi-perfection, has self-esteem issues and rarely speaks up. Perhaps symptomatically, many #70 songs are instrumental.

"Casanova 70," a track from Air's debut EP, *Premiers Symptômes*, takes its title from a 1965 Italian film about a man whose libido only kicks into gear when he's in mortal peril. With burbling analog synths and a sense of leisurely torpor, it feels like the theme song of *The Late Late Show*, circa the early Gerald Ford administration. You can practically feel the shag carpet growing under your feet. Phil Manzanera, the guitar wizard on the classic early Roxy Music releases and subsequent collaborations with Brian Eno, employs an array of effects to create a cathedral-like sense of space in "Europe 70-1," with nary a guitar in sight. Simple Minds might be best known for a certain moody anthem that plays over the closing credits of *The Breakfast Club*, but people forget the band worked in a host of styles. A lawnmower engine, for example, plays a critical rhythmic role on the turgid "Sound in 70 Cities."

> "Seventy ain't nothing but a damn number ... I'm writing and creating new stuff and putting together new different things. Trying to stay out there and roll with the punches. I ain't quit yet." —Bo Diddley

Reaching the age of 70 has to be a pretty freaky milestone. On the one hand, you're probably grateful for your longevity. At the same time, you're in shock and denial to look that bent numeral in the eye and call it your own. Wrote Charles Bukowski, "A man can go seventy years without a piece of ass, but he can die in a week without a bowel movement." Accordingly, it comes as no surprise to find a dearth of songs that address turning 70. Paul Simon did, albeit when he was in his 20s. Upon reaching that age in 2013, he confirmed to Maureen Dowd of *The New York Times* that his lyric for "Old Friends" ("How terribly strange to be 70") had been right on the money. "It *is* strange," he said. "It's not terrible, but it is strange." As for an actual #70 song that looks at 70, there's "What Will Be When You're Seventy," a crunchy screed by the Pack, the precursor to second-tier British goths Theatre of Hate.

---

[1] The next smallest weird number is 836, which, for 70, is like being a 1st grader with a sibling in grad school.

If 6 Was 9

"Look at Mother Nature on the run in the 1970s…"
—Neil Young, "After the Gold Rush"

By a wide margin, most #70 songs concern themselves with a certain decade associated with the smiley face, feathered hair, and double-digit inflation. "That '70s Song (Based on 'In the Street')" by Cheap Trick is an affectionate look back through rose-colored glasses at an era that was very good to the Trick. Now, Robin Zander and Co. know how to choose cover songs ("Ain't That a Shame" and "California Man" spring to mind), so one would be justified in having high hopes for their version of Big Star's anthemic "In the Street." And while the gorgeous melody and Beatles-eque harmonies would seem tailor-made for a pumped up, full-on Trick treatment, the track is overcooked. Worse, they baldly nick the bass part from Aerosmith's "Draw the Line," soften up the "wish we had a joint so bad" line as "wish we had a number so bad," and, worst of all, give *themselves* a shout-out, not Big Star—giving the whole thing a touch of travesty.

Maybe I'm getting a bit holier-than-thou. In "Losing My Edge," James Murphy of LCD Soundsystem obliterates the elitism of hardcore music fans, right down to their "white vinyl versions of every seminal Detroit techno hit." He also name-checks Yaz, who voiced similar sentiments about misplaced groupthink on the duo's breakthrough, *Upstairs at Eric's*. "Goodbye 70s" finds Alison Moyet at her gale-force best, raining down a molasses storm of good riddance on short-lived youth cults and fashion trends. Oddly enough, in league with the basic Yaz proposition is "'70s Music Must Die" by Lard, a side project of Jello Biafra and Al Jourgensen, which assailed "Bogus bands, plastic rock stars/Stupid clothes and the worst made cars."

Don't get me wrong; some songs do extol the era. Finding the intersection between punk and bubblegum, Jeff Dahl's "Circa 70" proudly proclaims the decade's rock gods ("Alice Cooper/David Bowie/Slade T. Rex") and ends with a snarled affirmation—"Still listenin' to that *shiiiiit*…"—sure to please any fan of *The Wire*'s Senator Clay Davis. The resilient Dahl, a veteran of the L.A. punk scene, former cohort of Stiv Bators, and one-time Angry Samoan, also sports a world-class 'fro in the "Waddy Wachtel/dude from the MC5" tradition.

THE VERDICT

In *Consider the Lobster and Other Essays*, David Foster Wallace wrote that in the 1970s, "[the] brave new individualism and sexual freedom deteriorate[d] into the joyless and anomic self-indulgence of the Me Generation." Mike Watt's "Against the '70s" has about as much warmth toward the era as Wallace did. Powered by Dave Grohl's explosive drumming and an Eddie Vedder vocal that makes better use of the man's low range than your average Pearl Jam howler, the song rages against the borrowed nostalgia endemic to musical elitists: "It's not reality. Just someone else's sentimentality."

Ball-Hog or Tugboat?, the source of "Against the 70's," sprawls over a host of styles, but this song feels like its calling card, representing a more straightforward style of rocking for Watt as well as evidence of his need to break from the past. When Watt came out with *Ball-Hog* in 1994, nearly 10 years had passed since the demise of the Minutemen, the vital San Pedro, Calif., trio he anchored with his athletic bass playing until the band's singer-guitarist, D. Boon, died in a van wreck. For his first solo record,

Watt enlisted a diverse roster of luminaries—including Henry Rollins, the Beastie Boys' Mike D and Ad-Rock, Frank Black, Sonic Youth, and Evan Dando—all of whom found something in the Minutemen to be inspired by.

It's not hard to see why. The Minutemen[2] possessed one of rock's rarest traits: uniqueness. Even with a seriously idiosyncratic set of influences (Wire, Funkadelic, Blue Öyster Cult), the band tossed out a lot of givens. The singer didn't sing so much as declaim; they cut about three-quarters from the running time of a typical rock song; they even dispensed with the expected subject matter (never wrote a single straight-up love song) and song-titling norms ("The Roar of the Masses Could Be Farts" is a classic.) Instead, drawing on and sticking to an M.O. devoid of ego and backstage riders, the Minutemen unleashed short bursts that were frantic, tight, and weirdly groovy. Drummer George Hurley deserves credit for fueling their astringent funk. "Against the 70's" is more traditional than a Minutemen song, and a lot less Minutemen-sounding than Watt's work with fIREHOSE—and I guess that's the point.

A SONG TITLE	AN ALBUM	A BAND	A LYRIC
"70's Scenario" – Hall & Oates (1974)	*70/90* – Tangerine Dream (1980)	Fela Kuti & Africa 70 – Afrobeat (1970s)	"70 years? It's neither one thing nor the other" – Julian Cope, "Elegant Chaos" (1984)

---

2   Another exceptional aspect of the Minutemen is that from their name onward, the trio had a marked predilection for numbers and numerical divisions. On their classic *Double Nickels on the Dime* is a profusion of number songs: "#1 Hit Song," "Two Beads at the End," "June 16," and "Take 5D." The album itself has "Maximum Speed 55" on the LP label, and throughout the band's discography are lyrics like "Number seven on the chump list" and "It's been 40 years/still a hostage," and "Let's say I got a number/that number is 50,000."

Seventy-one is a rare bird. You have to search hard to find meaningful instances of 71 in human enterprises, and when you find them, they seem trivial to the point of absurdity. But I call that fertile ground. To wit: There are 71 "cans" in "Yes We Can Can," a song by Allen Toussaint. (Go ahead and count them. I dare you.) In 1973 it provided the Pointer Sisters with their first hit, and was warbled memorably by a chorus of codgers in the 2004 doc *Young at Heart*.

Sexual slang aficionados use 71 to indicate a digitally enhanced 69. Aficionados of classic Swedish camping stoves never enter the woods without their trusty Primus No. 71. And aficionados of SR-71, a Baltimore alt-rock band of the '90s, know the band was named for a supersonic spy plane and not a classic Swedish camping stove. I should mention Robin Williams had a routine in which jihadists are rewarded in the afterlife not with "71 dark-haired virgins" but with "71 crystal-clear raisins."

The room-filling voice of disco diva Loleatta Holloway has powered a few decades' worth of dance hits, but it took the magic of sampling for her to reach the masses, most famously on "Good Vibrations," the 1991 chart topper by the artist formerly known as Marky Mark, which was built around a bit from Holloway's "Love Sensation." But back when Wahlberg favored Huggies over Calvins, Holloway was busy belting soul numbers in a brassy, sassy R&B style. "Rainbow '71" is her version of a Curtis Mayfield song that Gene Chandler, of "Duke of Earl" fame, turned into a hit—twice. (Presumably, Chandler's live take, titled "Rainbow '65," inspired Holloway's inclusion of '71 in her "Rainbow.")

Both singers modulate masterfully between restraint and release. Holloway takes her time getting there. The languid spoken-word intro ("You know, girls, it's *hard* to lose someone/especially if you lose that someone to someone you don't even like…") is the first of several asides to the girls; she doesn't begin to sing until about three minutes in. Along the way, the tempo picks up and eases back; there are sudden stops, the bottom drops out, and Holloway paints her pain on the music's expansive canvas, begging, then commanding, then howling the hurt away.

A different kind of drama permeates Bone Thugs-N-Harmony's "Down '71 (The Getaway)." The song begins in a noisy courtroom, where a judge (clearly the first cousin of the judge in "Living for the City") gleefully sentences Bizzy Bone to "death by electric chair"—but not until Bizzy, Wish, Flesh, and Layzie Bone tell their side of the story. Describing in trademark rapid-fire style how they killed rivals and police officers may not have been wise from a legal standpoint, but it sure makes for an entertaining day in court.

Monty Python ("Noooo-body expects the Spanish Inquisition!") and Mel Brooks ("You can't Torquemada anything") have guaranteed that my initial reaction to Electric Wizard's "Torquemada '71" is to snicker. But these Dorset, England, doom-metal masters make a fantastic noise, like Black Sabbath jamming in a room gradually filling up with syrup. Sabbath's own #71 song, "Weevil Woman '71," is comparatively light on its feet, a musical 400-yard dash in lead boots. And while this bonus track from the anniversary edition of *Master of Reality* could hardly be called a revelation, "Weevil Woman" (which predates ELO's "Evil Woman" by four years) retains the spirit and crunch of the seminal Ozzy years.

Heavy #71 stuff continues with the Oslo band Thulsa Doom, named for a powerful wizard who did battle with Conan the Barbarian and was portrayed by James Earl Jones in the film version. On "Generation 71," from the ingeniously titled *...And Then Take You to a Place Where Jars Are Kept*, Papa Doom unleashes his id and channels his inner berserker. It's primal sludgy stoner rock, proudly dunderheaded but with an absurdist edge: "She's got no ethnic friends/ but she knows someone with AIDS/ got coffee in a cup for days/ and it's the color beige."

Listening to "Conductor 71" by Fujiya & Miyagi begs the question: Am I listening to vintage Krautrock made by authentic Germans named Klaus or Horst, or is it the work of Steve, Matt, David, and Lee of Brighton? It sure burbles, crests, surges, and glides over one chord like the real thing. Makes sense, for it was Krautrock that brought Steve Lewis (Fujiya), David Best (Miyagi), Matt Hainsby (ampersand), and Lee Adams (kerning) together, and it is in Krautrock that they mostly abide. In the Krautrock version of Beatlemania, this is the overture.

"Octet '71" is an experimental work for chamber ensemble by Cornelius Cardew, an English composer who died in 1981 at the peak of his fame—some allege he was killed for his Marxist leanings. In the late '60s he led a group called the Scratch Orchestra that featured the youthful Brian Eno. Eno was profoundly influenced by Cardew's conceptual work "Paragraph 7," which calls for several vocalists to simultaneously sing brief sections of Confucius. Cardew's work has been performed by Sonic Youth, and he would seem ripe for a cinematic portrayal.

HIDDEN TRACK: Rosanne Cash's *Black Cadillac* ends with "0:71"—71 seconds of silence. In her autobiography, *Composed*, Cash explained: "To me, it was the only direct tribute track to my mother and father, both of whom had died at the age of seventy-one."

THE VERDICT

Magnolia Electric Co. was the project of Jason Molina, a songwriter heavily indebted to Neil Young—both the bare-bones, rockin' Neil and the country-rooted, twangin' Neil. "Texas 71," from the four-disc, fans-only *Sojourner*, recalls Country Neil, with Molina's plaintive vocal augmented by Mike Brenner's mournful peals of pedal steel and Michael Kapinus's lulling organ. The suggestion is of Young's "Helpless," only instead of north Ontario, the song evokes a dusty Texas highway (presumably the one stretching 250-some miles from Brady to Blessing).

And the year '71—when Young and the U.S. both found themselves at a listless juncture—is in there too. The verses mix ache and resolve, the chorus providing the uplift. Molina is joined in rough-hewn harmony by his bandmates, and they enunciate the title phrase for all it's worth, cushioning the hurt in Molina's voice.

Beginning in 1996 as Songs: Ohia, Molina was extraordinarily prolific, but by 2007, when "Texas 71" was released, he was deep into his struggle with alcoholism, which had plagued him for years. During a 2009 tour of Europe, when he began to fall apart, Molina finally admitted his problem. For the next four years, he made sporadic efforts to clean up. Jason Groth, who played lead guitar for Magnolia Electric Co., told *INDY Week* that Molina's life became nothing short of a battle with himself. "[I]t got so out of hand that he was unwilling, or maybe unable, to decide every day when he woke up to not take a first drink and let it take over his life."

It did. Molina's organs gave out in 2013. He was 39. Groth and some members of Magnolia Electric Co. got together in early 2014 and played four concerts. Michael Taylor of Hiss Golden Messenger joined them to sing Molina's parts a few last times before laying the songs to rest for good.

A SONG TITLE	AN ALBUM	A BAND	A LYRIC
"71" – Ben Webster (2005)	*Cluster '71* – Cluster (1971)	The 71's – Houston, Texas, rock (2006–2010s)	"Heaven's number is 71" – Owen Pallett, "Lewis's Song" (2009)

72

On "U.S. Blues," Jerry Garcia says his pulse "stays 72, come shine or rain," but it's clear from his numerous near-death experiences that his pulse was in fact a very changeable thing. This doesn't change the fact that the foregoing might be the most distinctive #72 lyric going. Bolstering the Dead's seventy-two-vian bona fides is *Europe '72*.[1] This sprawling chronicle of the band's fabled improvisatory art in full bloom[2] was one of the first three-LP (hence double-gatefold) packages, opening up new frontiers in doobie rolling, much as the aluminum bat transformed Little League baseball two years later. On that legendary European tour, the Dead played a pair of gigs at the Tivoli Concert Hall in Copenhagen, the Danish capital city whose Latin name, Hafnia, gives the chemical element hafnium, otherwise known as atomic number 72, its name.

The Dead may have been riding high in '72, but the year was fractious and ugly by any measure: the massacre at the Munich Olympics, Vietnam, and Nixon in a landslide. In "Back in '72," a rousing road song that doubles as a sober critique of the peculiar excess of the time, Bob Seger mentions the newly re-elected president ("Tricky Dick he played it slick") and laments a year in which "we learned nothin' new." In an entirely sillier vein, Dickie Goodman's "Convention '72" exemplifies a novelty-song trope that had a few decades' worth of traction: stitching lines from popular hits into goofy narratives. Goodman, the Weird Al of his day, scored a string of hits of a similar topical bent, including "Superfly Meets Shaft" and "Mr. Jaws."

"72 and sunny" has come to embody the ideal meteorological experience on planet Earth while also serving as the name of an L.A. ad agency, an album by Uncle Kracker, and a trifle on the *WALL-E* soundtrack. On the non-sunny side, "Emergency 72," by the folk-minded British duo Turin Brakes, takes its title from emergency kits designed to sustain survivors for 72 hours following a disaster. And Bobby Hebb, who wrote the hit "Sunny," died at 72.

THE WEED CONNECTION

In the Book of Genesis, the postdiluvian world comprises 70 nations, each founded by one of Noah's offspring. In the Jah-praising "72 Nations," which stretches out on a single chord like a blissed-out "Papa Was a Rolling Stone," Dadawah, better known as Ras Michael, adds two extra nations. "72 Kilos" by Keziah Jones refers to the potent marijuana native to Nigeria, which, according to growinghelp.com, produces "avalanches of hard, fist-sized buds that bristle with spiky pistils and glisten with a slick coating of trichomes."

---

1 For the Deadhead who can't say no, the entire European tour was released in a 73-CD box set in 2008. Sales were brisk, yet on one online Deadhead forum, a listener admitted to some extremely slight misgivings: "I noticed some hiss and warble on the end of El Paso (disc 1, track 7) just after the song ended." Presumably the rest of it was rockin'!

2 Guterman and O'Donnell's hilarious demolition of Europe '72 in *The Worst Rock n' Roll Records of All Time* ("It's just solo, solo, puff, snort, solo, solo") makes a weird kind of sense to me, even though I've always dug the LP deeply.

If 6 Was 9

THE VERDICT

The stretch of ruinous asphalt known as U.S. Highway 72, whose title number appears prominently on the cover of the Drive-By Truckers' *Southern Rock Opera*, has the distinction of being the only U.S. highway that begins and ends in the same state (while passing through other states). In the two sinister chords that open "72 (This Highway's Mean)," the Truckers conjure the road's desolation. Mike Cooley, whose vocals usually possess an audible snarl, is understated yet urgent; he doesn't want us to miss a word.

As one of the band's two primary songwriters, Cooley is less prolific than Patterson Hood, but he's a skilled and distinctive lyricist and writer of some of the band's most enduring songs. *Southern Rock Opera*, a largely self-financed, double-sided phantasmagoria, imagines the story of Lynyrd Skynyrd within the larger context of what Hood termed "the duality of the Southern thing." On a record rife with the ghosts of Ronnie Van Zandt and George Wallace, "72 (This Highway's Mean)" is more concerned with place than with character. It establishes a setting for the sprawling story that follows. Cooley sings in a tar-deep baritone, never identifying the road by its number but assuring us that it's old, mean, and dusty and that we should avoid it at our peril.

In a 2013 email, Patterson Hood said, "I know he wrote it because it's the only road out of town to anywhere else in my hometown (at least to anywhere we'd have wanted to go) and was the road to Memphis, which was the Promised Land to us back in the day (until we made the mistake of packing our shit and driving up 72 to live there, for several miserable months).

"It was (in our youth) a windy and very dangerous two-lane speed trap and death trap with bad curves and horrible accidents at stunning frequency. (Now it's all four laned and a much tamer and shorter drive). I'm sure Cooley is aware of the other song mythologies of highways (61, Lost, etc.) but I don't know how much that played into his decisions or its creation. I doubt he'll tell either, but you never know."

A SONG TITLE	AN ALBUM	A BAND	A LYRIC
"72" – James (2008)	*72* – Leadheart Deadbird, Baltimore hip-hop (2010s)	The Delta 72 – Washington, D.C. alt-rock (1990s–2000s)	"He remembers Roxy Music in seventy-two" – Belle & Sebastian, "Me and the Major" (1996)

73

The Blues Image were a lot bluesier than "Ride Captain Ride" might indicate. Jimi Hendrix sang their praises, and he knew a thing or two about the blues. But it's solely by that maritime-utopian foot-tapper—which nearly topped the U.S. charts in July 1970, two months before the death of Hendrix—that the band is remembered.[1] The song made a strong impression with an opening line whose sheer specificity seemed to suggest an almost historical truth: "Seventy-three men sailed up from the San Francisco Bay." Many of us wondered about those 73 men: Were they real? How does one secure passage on a "mystery ship"? But, according to Mike Pinera, the song's lyricist, he summoned the oddly numbered crew purely from his imagination during a recording session.

After falsely assuring his producer that he had something, Pinera went and meditated awhile. With an unfettered mind, he cast his gaze on the 73 keys of his Fender Rhodes Mark I Stage Piano. "So I say, 'Okay, I need a first word.' And what came into my head was 73. I liked the rhythm, and I went, '73 men sailed up, from the San Francisco Bay.' The song sort of just wrote itself."

The songs of Richard D. James, aka Aphex Twin, on the other hand, could never write themselves. The cunning hand of a sometimes demented-seeming musical genius makes its presence felt, whether in a latticework of jarring rhythms or in ethereal flotillas of synth ambience. "73 Yips" is of the former style, and its twitchy unsettledness seems an apt musical evocation of "the yips," a nervous condition that renders athletes unable to do the most basic things, like throw a ball to first base or hold a putter steady.[2] Given James's penchant for inscrutable titles (his most beautiful song might well be "Flim"), it is hard to say with certainty that he meant yips in the

MATCH THE #73 LYRIC TO ITS SOURCE

1) "Seventy-three inches of all-black everything, laid out like a ramp"

2) "The scabs outside still laughed at their spree/ And the children that died there were seventy-three"

3) "Seventy-three was a jamboree/ We were the dudes and the dudes were we"

4) "I was followed home by a 73"

5) "The best number is **73**. Why? 73 is the 21st prime number. Its mirror (37) is the 12th, and its mirror (21) is the product of multiplying 7 and 3"

A) Sheldon Cooper, *The Big Bang Theory*

B) Robyn Hitchcock, "De Chirico Street"

C) *Esquire* feature on Jay-Z

D) Woody Guthrie, "1913 Massacre"

E) Mott the Hoople, "The Saturday Gigs"

Answers: 1-C, 2-D, 3-E, 4-B, 5-A

---

[1] The band also never receives credit for its role in ushering in, albeit unconsciously, the great era of what might be called "captain-philia." As the '60s slipped uneasily into the new decade, you couldn't turn the radio dial without running smack into a captain. There was Grand Funk's "I'm Your Captain," Neil Diamond's "Captain Sunshine," the Doobie Bros' *The Captain and Me*, Billy Joel's "Captain Jack," and Elton John's *Captain Fantastic*, not to mention the Captain and Tennille. (And Captain Beyond, a great stealth supergroup, formed in 1971). Perhaps this profusion of captains, along with what would now be termed the re-trending of Jesus Christ, indicates that young people were looking for a steady, manly force to guide them.

[2] The yips are not usually a factor in competitive gymnastics, certainly not for Nadia Comaneci, who wore the No. 73 on her narrow back when she earned a perfect 10 score on the uneven parallel bars at the 1976 Montreal Olympics.

sporting sense. Indeed, James would be an unlikely sports enthusiast. However, he does hail from Cornwall, and Cornwall is a stronghold of cricket, a sport in which the yips often afflicts "left-arm spinners." Oddly enough, both James and Chuck Knoblauch, who had one of the more famous cases of the yips, made their debuts in 1991: James with *Analogue Bubblebath*, on the label he co-owned, Rephlex Records; Knoblauch as second baseman for the Minnesota Twins.

When he later joined the Yankees, Knoblauch would stride to home plate to the strains of the Beastie Boys' "Intergalactic," so you'd best believe he had no interest in offerings like Tender Trap's "Face of 73." This brisk squirt of twee pop does not name the visage in question, although if there *was* a face of '73, it might have belonged to Cheryl Tiegs (who hails from Minnesota and is the mother of twins). Tender Trap was pretty much a vehicle for Amelia Fletcher, who previously fronted Tallulah Gosh and Heavenly in the late '80s. In "Face of 73" she takes aim at the artifice of, and the public's hunger for, en vogue images that don't last. One legacy of 1973 that *has* lasted is the American League's adoption of the designated hitter, a position Chuck Knoblauch played when his yips became a serious liability.

THE VERDICT

Songs called "Think" unite the Queen of Soul and Soul Brother No. 1. Aretha Franklin co-wrote hers with her then-husband, Ted White, and this 1968 liberation anthem followed musically and thematically from her signature hit, "Respect," of the previous year. James Brown's "Think," which he first performed with His Fabulous Flames in 1960, is by Lowman Pauling, a member of R&B archetypes the 5 Royales, who had their own hit with it.[3] Aretha's version of events is a pointed finger in the face. Brown's "Think" is more of a sweeping gesture toward the horizon: think about the good things, the sacrifices I've made. He even implores his love to consider the bad things he *tried* not to do. I can't imagine Aretha making the same statement. In any case, the song went through several transformations, from the loose-hipped blues swing of the Royales into J.B. & Co.'s rhythmic tour de force, which has the urgency of a "Reveille" call. By the time Brown revisited the song and re-dubbed it "Think '73," his tempos had shifted toward a deeper shade of funk.

Think what you're doing to me—it's the guilt card played as an accusation, and it's devastating. A songwriter who wields this imperative gets a platform to say who's thinking and who is not, to list all the things that should be thought about, reconsidered and regretted. Brown's remake surely warrants a retitling. The message is unchanged, but the difference in feel is so marked as to render it into a new song, tailor-made for the era of wide ties, elephant bells, and orb-shaped 'fros.

A SONG TITLE	AN ALBUM	A BAND	A LYRIC
"73 in '83" – The Legend (1983)	*Seventy Three* – Seventy Three (2004)	Prefuse 73 – Atlanta, Ga., Electronic/rap (2000s–2010s)	"All night it's a balmy 73 degrees in Fog City" – Cheap Trick, "On the Radio" (1978)

---

3   Pauling penned "Dedicated to the One I Love," was a wickedly good and influential guitar player, and never got the credit for any of it. When he died, he was employed as a custodian in a Brooklyn synagogue. The year was 1973.

74

The trend toward remakes, remixes, and other ways of reworking old material has provided songwriters with an enduring trope, wherein a number indicating a year is tacked on to the end of a song name. Despite exceptions like "Bo Diddley '69," the number usually goes unmentioned in the lyrics. Alice Cooper's "Teenage Lament '74" is no remake, but its evocative title confers a quasi-historicity on this anthemic track, which sports backup vocals by Liza Minnelli and a passel of Pointer Sisters. Like most of Alice's mid-'70s singles, it reached the Top 20. (So did "Energy Crisis '74" by novelty song-meister Dickie Goodman, although, as with Cooper, his best work was behind him.)

While Cooper was notorious for beheading chickens, Stone the Crows, a Glaswegian pub band led by the brassy-voiced Maggie Bell, had no beef with birds—they were named for an expression of surprise familiar to the United Kingdom. "Love 74," written by keyboardist John McGinnis, morphs through a Zeppelin-esque[1] groove, some meandering piano reminiscent of Traffic, and a few David Gilmour-esque slide guitar bursts, courtesy of Les Harvey, who died onstage in Swansea in 1972. "Love 74" lacks an apostrophe, so the meaning of 74 is unclear.

No such ambiguity exists in "74 Is the New 24" by Giorgio Moroder, a pioneering figure in electronic dance music who, like Aphex Twin, reappeared in late 2014 after a long silence. Moroder, who has topped this list by proxy for co-writing "The Number One Song in Heaven" with the Mael brothers, returned to the world stage not with a blimp, but with the release of a new song, a seizure-inducing video, and an announcement that a star-studded full-length release was forthcoming. The news was met with quasi-religious zeal. With a worldwide audience primed by Daft Punk's tribute, "Giorgio By Moroder," on the previous year's *Random Access Memories*, the timing was impeccable. Reaction on SoundCloud, where the song was posted, ranged from sheer awe to amazement that a man well past retirement age could produce such "sick" beats.

"74 Is the New 24" sounds exactly how a Moroder song should sound. Prominently featuring the click track he made famous, along with the oscillating synth tones of his *Midnight Express* soundtrack, the song robotically delivers the defiant message of its title over an irresistible dance-floor groove. Viva Giorgio!

THE VERDICT

The Connells shared the sound, spirit, and Southern origins of their contemporaries R.E.M., only they were a lot less mysterious. Still, their lean, melodic rock songs were embraced by the jangle-minded, college-rock-loving youth of the day—myself included—

---

[1] Stone the Crows' manager and eventual producer was Peter Grant, the gargantuan, Mephisthophelean force behind Led Zeppelin.

and the band became a dependable live draw. Or, as Steve Gottlieb, the head of the Connells' notoriously tightfisted record label, TVT, put it, "a musician's band that plays for their fans." Which of course no band wants to be.

The band's crowning glory, 1987's *Boylan Heights*, was every bit the equal of the record R.E.M. put out that year (*Document*). Yet in a recent, slightly guilt-ridden appreciation of the Connells' signature song, a blogger at *The Guardian* called them "an inoffensive guitar band who wear inoffensive shirts."

Shirts aside, the haunting acoustic intimacy of "74-75" was in marked contrast to the bombast that was the prevailing aesthetic during the mid-'90s. Coupling touches of Celtic folk with transporting, monastic-style background vocals in the chorus, the song was a hit—or, I should say, a hit in Europe.[2] Back home, where it mattered most, the success was more modest. Even a popular and poignant MTV video did not translate into widespread album sales. Already weary from years of touring and record company angst, the band soldiered on, but there would be no capitalizing on the promise of "74-75."

Speaking with David Menconi of Raleigh's *News & Observer* in 2007, Mike Connell said the song was not about specific years, but rather "people reaching a crossover point of no return in their lives." So what about those crooked digits? "The only reason Connell picked the numbers in the title," wrote Menconi, "was that they sang well within the song's meter."

Just as I suspected…

A SONG TITLE	AN ALBUM	A BAND	A LYRIC
"74 Cuts, 74 Scars" – Centro-Matic (2007)	*A Toot and a Snore in '74 [bootleg]* – John Lennon, Paul McCartney, Stevie Wonder, Harry Nilsson, et al. (1974)	74th Street Band – U.S., roots rock (2000s)	"Took the Dodge Dart, a '74" – A Tribe Called Quest, "I Left My Wallet in El Segundo" (1990)

---

2   It was a chart-topper in Italy, Sweden, Israel, and Norway, where radio was especially friendly toward the melodic alternative rock of mid-level U.S. bands in the 1990s. The song that supplanted "74-75" at the top of the Norway charts was "You Suck" by the folk-pop duo the Murmurs—hardly a radio staple stateside. The decade also saw Maria McKee and the Bloodhound Gang reach the No. 1 spot in the Land of the Midnight Sun.

75

After the unexpected success of "Tequila," a hastily written B-side that went to No. 1 in early 1958, the Champs quite naturally tossed back a few more shots. First came "Too Much Tequila," then "Tequila Twist," both with diminishing returns. The band veered into "Limbo Rock" territory before returning to the subject of drinking in 1965, although this time the Champs opted for something a little more upscale. "French '75" refers to a Champagne-and-gin cocktail (a favorite of Ernest Hemingway) named for a French artillery gun. A tidy twanger in the Ventures vein, the song lacked fizz and bubbled under.

Tequila—the drink, that is—flourished, aided in no small part by the Rolling Stones, whose passionate endorsement during 1972's so-called Cocaine and Tequila Sunrise Tour helped turn it into one of rock's premier libations. And they never even sang about it. Eventually the Stones would also spawn the Brian Jonestown Massacre, whose air of bacchanalia did honor to the London band's notorious decadence. BJM maintained its Stones fixation on album No. 2, *Their Satanic Majesties Second Request*, a trippy sprawl that included "Miss June '75." Written by third banana songwriter Matt Hollywood, who left the band after an onstage melee captured in the engrossing film *Dig!*, it's a lot more sincere than its *Playboy* centerfold-sounding title would indicate. The resolutely simple chord progression and confessional POV, not to mention its culminating act of psychedelic oral pleasure, seem weirdly, cosmically sincere.

The songs of John Darnielle are suffused with their own brand of weird, cosmic sincerity. At first blush, "02-75" looks like its title refers to a date, but the lack of apostrophe is telling and so is the dash. Turns out to be a post office box. Though it hasn't been on an official release, "02-75" (pronounced "oh two dash seven five") has remained in the Mountain Goats' set list since 1996. During a 2008 Daytrotter session, Darnielle introduced it as "a love song for my wife before I knew she was my wife." Darnielle's characters usually exist in less hopeful situations. Even his love songs can get surrealistic, like the one where he compares his heart to a high-pressure sterilization chamber. But here the emotion is pared down to its essence: "You are my best friend and I have always known you."

THE VERDICT

Just as John Darnielle knew his wife, Lalitree, was out there in the world, in the summer of 1969, Wally Shoup and his pals also knew something significant was out there: a psychedelic rock record to top them all, and they pursued it as if it were the lost chord. Once they managed to get their hands on the mythic platter—the band and LP were both titled *Touch*—and having gotten themselves "properly prepared" for the journey it promised, they let it rip, ready to have their minds blown.

And it was good—very good even—but not quite the ultimate. At least not until they reached the final track…

"Then came the clarion, Morse-code-like beginning to the final cut, 'Seventy-Five,'" wrote Shoup, an accomplished alto sax player, in 2004, "and, for the next twelve minutes, we were transfixed, buffeted and hurled through a sonic universe that shook us to our boots."

The mastermind behind Touch was Don Gallucci, an original member of the Kingsmen who, at age 15, played the iconic organ riff on "Louie Louie." Unable to tour with the Kingsmen because of his age, Gallucci eventually put together a band of his own and scored a minor hit with the Beach Boys-lite of "I Could Be So Good to You" in 1967. Not long afterward, with the help of mind-expanding psychedelics, Gallucci read the ancient Venusian rune writing on the wall and concluded that sunshine pop was old hat. Armed with strong hallucinogens, he and his fellow astral travelers set out to make a musical statement that reflected their consciousness-raising visions. Rehearsing in a castle in the Hollywood Hills, the 19-year-old Gallucci and his bandmates were sincere in proselytizing on behalf of the drug; they truly wanted their music to get people tripping. Gallucci called their work "a spiritual quest put to music."

During the recording sessions, an aura arose around the project. Grace Slick and Jimi Hendrix, among others, dropped by to take in the vibes. Although making a record under the influence of strong drugs can be disastrous, *Touch* is marked by rock-solid performances, arrangements, and production. Its undisputed highlight is the 12 minutes of "Seventy-Five." With prominent church organ, a free jazz interlude, ethereal, high-pitched vocals (one wag on YouTube likened "Seventy Five" to Rush meeting the Supremes) and astounding culminating five-minute build, it lives up to the well-worn "epic" tag. The song totally anticipates Yes, from the caustic keyboard at the end, which prefigures the "Roundabout" solo, to quiet sections featuring the ethereal counter-tenor of guitarist Joey Newman.

Yet Touch has zero recognition among any but the most committed of music buffs, and while that's sad, it's also understandable. Don Gallucci[1] decided the music was too complex to be reproduced in a live setting, so a planned tour with the Moody Blues was scrapped. The album was a flop and the band broke up soon thereafter. We can thank the Internet for providing access to this lost gem, so get thee to YouTube and, in the words of a different Touch song, "slowly pull back your mind petals and listen listen listen."

A SONG TITLE	AN ALBUM	A BAND	A LYRIC
"Melochord Seventy Five" – Stereolab (1995)	*Neu!'75* – Neu (1975)	75 Dollar Bill – Brooklyn, N.Y., lo-fi/microtonal (2010s)	"Look to the summer of '75" – Jefferson Starship, "Ride the Tiger" (1974)

---

1 The Zelig-like Gallucci went on to produce the incendiary *Fun House* by the Stooges, making him a key player in primal '60s rock, proto-punk, and prog rock.

Moe: "I say, Jasper, what comes after 75?"
Larry: "76."
Moe: "That's the spirit."

"The Spirit of '76" is an iconic painting by Archibald MacNeal Willard, depicting three musicians—two drummers and a fife player—responding to the muse as they stroll through a battlefield while behind them fly the Stars and Stripes. Also known as "Yankee Doodle Dandy," the work was displayed at the Centennial International Exhibition of 1876, the precursor of the World's Fair, which was held, appropriately enough, in Philadelphia.[1] The City of Brotherly Love, as we will soon see, makes an appearance in nearly every notable #76 song. And the title, if not the spirit, of Willard's painting has inspired a handful of songwriters—unfortunately not always in edifying ways.

Todd Rundgren's "Spirit of '76" went wisely unreleased until it ended up on the tail end of a massive collection of demos and outtakes by the Philly-born neo-soul polymath. This wah-wah-drenched pastiche incorporates "The Star-Spangled Banner" and "Dixie" trilled on a fake fife and is as cringe-worthy as anything ever done by the man known by his fans simply as Todd. The Alarm's "Spirit of 76," a reminiscence about bygone friends and lovers that's heavy on gleaming chords and overwrought vocals, is several munitions shy of "68 Guns." Better, but not by much, is "A Tribute to the Punks of '76," a medley by the Friendly Hopefuls, which mocked the chart-topper of the (1981) moment, "Stars on '45."

No doubt, a red-white-and-blue spirit courses through songs that incorporate 76 in their titles. *The Music Man*'s "Seventy-Six Trombones" is a rousing salute to orchestral excess that remains a popular standard with the Boston Pops and its ilk. The Philadelphia-centric hip-hop duo G. Love and Special Sauce saluted the big highway that runs through the city in "I-76." Like the Alarm's "Spirit," the song is a fond remembrance of a bygone era, but instead of lingering over the dude who ended up in jail or the girl whose smile was brighter than the sun, G. Love and company name-check Mo Cheeks and Moses Malone. And they offer practical advice: If you're in a hurry, there is no substitute for I-76, they counsel, "unless of course you wanna take the scenic view/ Then the East or West River Drive is just right for you."

---

1 Also on view at the event: Bell's telephone, the first typewriter, and Heinz ketchup.

If 6 Was 9

THE VERDICT

Pop music has a long and rich history of novelty songs, yet most bands that desire to be taken seriously tend to avoid outright humor. Black humor has its place, as does cheeky humor and deadpan irony, but abject jokiness, straight-up absurdity in the name of big yuks, has been the province of a select few. Frank Zappa, who famously asked whether humor belongs in music, certainly comes to mind, but he balanced his jokiness—equal parts sophomoric bathroom humor and biting social commentary—with music that was ambitious and demanding.

Dean and Gene Ween[2] were also proud, talented, ambitious goofballs with serious chops. Their first few releases as Ween fully embraced anarchic silliness, splattering multiple genres against the wall with dexterity and abandon. Through it all, their mastery of genre was clear, whether in Queen-like intricacies, noise experiments, dead-on country-and-western, bedroom beatbox, or helium-treated sunshine pop.

"Freedom of '76" comes from Ween's undisputed masterpiece, *Chocolate and Cheese*, a record that has earned its own entry in the 33 1/3 series[3] and is beloved to the hardest of hardcore Ween fans (Weenists?). On the duo's first releases, Freeman's voice was often manipulated, sped up, slowed down, texture-treated. But his earnestly soulful, smile-inducing falsetto vocal on "Freedom of '76" shows a seriousness of intent that is largely absent from the early stuff.

In a YouTube video, Dean Ween explains, "It's unique among Ween songs in that it was our first song where we used a lot of major seventh and minor seventh chords that were shown to me by my friend Ed Wilson. They are not chords you would normally associate with rock 'n' roll music. It's more jazz you would think of with these chords."

The result is a spot-on tribute to a somewhat mythical City of Brotherly Love. It being Ween, you expect something strange or tongue-in-cheek to enter the proceedings, but nope, this is Ween straight, and the effect is not unlike seeing Bill Murray in *Lost in Translation*. The level of soul on display catches you off guard. Gone for the moment are the household inhalants that the Weens indulged in during the *Chocolate and Cheese* sessions. This one's from the heart, a languid arrow to pierce the Philly Soul sweet spot.

A SONG TITLE	AN ALBUM	A BAND	A LYRIC
"Seventy-Six" – The Sadies (1998)	76 – Armin van Buuren (2003)	Charanga 76 – Latin, (1970s–1980s)	"Well, by '76 we'll be AOK" – Donald Fagen, "IGY" (1982)

FINAL NOTES:

The chord sequence that cycles through "Freedom of '76" culminates in the so-called Hendrix chord, a dominant seventh sharp ninth that Jimi used most conspicuously in "Purple Haze."

---

2 The aliases of Mickey Melchiondo/Dean and Aaron Freeman/Gene.
3 As described in Hank Shteamer's book, the song's rich vocal harmonies were built up in a tedious process inspired by Todd Rundgren's painstaking, all vocal *A Cappella*. In the book, Melchiondo clarifies the song's *Mannequin* reference: "*Mannequin* was actually filmed in Wanamaker's in Philly, which was a classic twentieth-century department store."

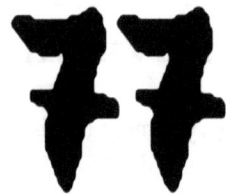

Wikipedia alerts us to the fact that 77 is extremely tricky to pronounce in Swedish, which is why *sjuttiosju* became a military password at the Norway-Sweden border during the Second World War. Perhaps tellingly, of the smattering of 77-titled songs in existence, none include an enunciation of the 22nd discrete semiprime in Swedish. It does sound great in English. It's got a cool five-beat cadence; it adds flash to the debut by Talking Heads, and it stands out like twin lightning bolts. The early-'60s detective sitcom *77 Sunset Strip* leveraged 77's sibilant qualities. Yet songs that actually say the number are rare. The instrumental "Mondo 77" is a catchy bit of new-millennium-era electronica by Looper, a mostly one-man show led by Stuart David, the original keyboardist for Belle & Sebastian. With a minimal, bleepy dance groove adding an urgent pulse, the track was frequently tapped for TV shows and ads.

Osaka, Japan's avant noise-meisters the Boredoms, led by vocalist/provocateur Eye Yamatsuka, have a thing for 77, with a pair of #77-titled tracks to their name. "G.I.L. '77" comes from *Soul Discharge*, a paragon of cacophony that aficionados of the genre hold in high regard. "77," from a 7-themed remix album that also features "7," "777," and "7777," is hypnotic, its central arpeggio taking on majesty as it cycles and pulses through rapid, Möbius strip-like repetition. Taken together, the two present a pair of extremes roughly equivalent to Brian Eno's "Baby's on Fire" considered next to his *Music for Airports*.

The hardcore punk of the early 1980s that inspired the Boredoms owes much to the Bad Brains, the Washington, D.C.-based group named after a Ramones song. The group rewrote the rules on hardcore, tricking out their spring-loaded crunchy speed riffs with elements of reggae and classic rock, not to mention a crisp, bright sound (their first proper release was produced by Ric Ocasek) that was far from the order of the day.

After the onslaught of several pummeling, riff-based explosions on the band's most lauded release, *I Against I*, "Secret 77" acts as a caesura in the maelstrom. Weirdly, it almost sounds like a metal song as done by the Police: That flanged guitar has more than a touch of Andy Summers, and the halting cadence of Earl Hudson's heavily gated drums suggest an unconscious channeling of "Behind My Camel" from *Zenyatta Mondatta*. Thankfully, the reggae croon of lead singer H.R. (it stands for "human rights") is not Sting-like by any stretch, but it's a world away from the barely suppressed venom of the tracks that surround it. Along with some spirited doggerel, the lyrics point to righteousness and oppression, but the title enigma is never resolved.

No such enigma is found in "England 77" by the ersatz punk group Horrorcomic. Starting and ending with a quote from Elgar's "Pomp and Circumstance," Britain's unofficial English national anthem, the song has the Rotten-esque upward sneer and the proud, bratty spirit of real punk. It also ranks as one of the first songs to look back on that pivotal year and declare its historical value. Comprising members of a glam-leaning outfit called Crackers, including original Sweet drummer Frank Torpey, Horrorcomic released just three singles. "England 77" was a B-side, but it's by far their finest moment.

If 6 Was 9

THE VERDICT

I must admit I was surprised to learn that the nightclubs of Mumbai in the 1970s thrummed with the sound of trippy, humid, psychedelic jams played by bands that made liberal use of effects pedals, synthesizers, and smoke-ables. My ignorance is excusable though; of the dozens of acts on the scene—like Velvette Fogg and the Savages—only one of them actually made a record. Its sheer rarity and unexpected provenance has turned that LP, Atomic Forest's *Obsession '77*, into an extremely valuable piece of vinyl. As of this writing, one copy is for sale on the online marketplace Discogs—for $5,000.

Atomic Forest was a four-piece led by the volatile, drug-addled bass player Keith Kanga, who financed the record with proceeds from his grandmother's brothel. What Kanga shared with his bandmates was a deep love for Western music. Vocalist Madhukar Chandra Dhas, for example, known as Madoo, had an epiphany as a child upon hearing "Twist and Shout." Madoo, one of the preeminent vocalists in that rarified milieu, ultimately split from the band, disgusted by Kanga's unscrupulousness. Madoo may not have made it onto the record, but he did get to jam with Jimmy Page and Robert Plant in 1973, when they dropped by the rundown Slip Disc club and got royally, memorably rock-star drunk.

The one original song on Atomic Forest's historical platter, which also featured a cover of Jethro Tull's "Locomotive Breath," is its fuzzed-drenched title track. Based on "Melting Pot" by Booker T. and the M.G.'s, and featuring scintillating solos by guitarist Abraham Mammen, it sizzles like a vindaloo-slathered funk grenade. The song is divided into two sections: "Obsession '77' (fast)" and "Obsession '77' (slow)." Take your pick—they're both at the same tempo.

A SONG TITLE	AN ALBUM	A BAND	A LYRIC
"Kiss in 77" – James Brown (1977)	*Talking Heads: 77* – Talking Heads (1977)	The 77s – Sacramento, Ca., rock (1979–2010s)	"September '77/Port Elizabeth weather fine" – Peter Gabriel, "Biko" (1980)

LAST WORDS:

In an attempt to gain entry into the *Guinness Book of World Records* for longest song title, the B-side of Fairport Convention's "Now Be Thankful" was dubbed "Sir B. McKenzie's Daughter's Lament for the 77th Mounted Lancer's Retreat From the Straits of Loch Knombe, In the Year of Our Lord 1727, On the Occasion of the Announcement of Her Marriage to the Laird of Kinleakie." It failed.

78

Like 33 and 45 before it, 78 holds an important spot in modern music by virtue of format. The 78 RPM record was fragile, noisy, and inefficient, yet it served. At first 78s were simply called records, as in gramophone records; later they were shellacs, for the material that coated them. It was only after World War II that the retronym 78 came into use, to distinguish them from the 33 1/3- and 45-rpm discs that had become standard. The first million-selling record was a 78,[1] and though they began to be phased out in the '50s, 78s have endured in the iconography of popular song as signifiers of bygone days. "Worn-out 78s" appear in the Kinks' "One of the Survivors," for example, and there are "Ink Spots 78s" in "Beecharmer," Nellie McKay's duet with Cyndi Lauper.

The brevity and surface noise of 78s well suit Stiff Little Fingers, a foursome out of Belfast whose "78RPM" is a brisk, barbed riposte to those who declared punk dead after the official demise of the Pistols (in early '78). It puns on 78 revolutions and lambastes the British government for condoning "acceptable levels" of violence in Northern Ireland. Most of all it rocks with passion and precision, incorporating the pummeling bubblegum changes of the Ramones while remaining light on its feet.

The unmistakable surface noise of an old shellac opens "78 Stone Wobble" by Gomez, an English foursome with a Spanish name who won England's Mercury Prize in 1998.[2] A standout from the band's award-winning LP, *Bring It On*, the song oscillates between Beck-ian acoustic hip-hop and a narcotic Beatle-esque chorus ("Open-hearted surgery never works") that I've pondered more than once.

On its second release, the pioneering German band Neu! recorded "Super" at three different speeds, not for conceptual reasons but to fill out the record that the duo lacked the money to finish. Thus "Super 78." But it isn't like Van Morrison's contract-killing drivel; in fact, the experimental second side of this 1973 record has been widely heralded as a precursor of the remix album, a form that grew ubiquitous a decade later. "Super 78" gallops along at hyper-speed at the same approximate clip as Ministry's "Jesus Built My Hotrod." It's entirely possible that the earlier track influenced the later one.

ON EXTENDED REPEAT IN HELL: "Summer of '78," Barry Manilow's salute to his soft-rock touchstones, from Little River Band to Firefall to "Sometimes When We Touch."

---

1 Enrico Caruso's 1904 recording of the aria "Vesti la giubba" from *Pagliacci*.
2 England's Mercury Prize, awarded since 1992, has had a seemingly career-crippling effect on many a hot UK act. Besides Gomez, M People, Roni Size, Dizzee Rascal, the Klaxons, and Elbow are among the honorees who failed to capitalize significantly upon earning the Mercury.

If 6 Was 9

THE VERDICT

Continuing the Velvet Underground after the departure of Lou Reed in 1971 was, in retrospect, a pretty crazy idea—sort of like keeping the Doors going without Jim Morrison, which the remaining Doors did anyway. But even after the departure of founding guitarist Sterling Morrison and drummer Maureen Tucker, the remaining member, Doug Yule, soldiered on. Instead of finding some sullen, artsy-fartsy would-be Lou type, or at least someone with a New York accent, Yule brought in Willie "Loco" Alexander, a stalwart of the Boston rock scene who was Yule's former bandmate in a group called Glass Menagerie. It somehow added insult to injury that this Velvet Underground in name only would fill its depleted ranks with a Bostonian. But they gave it a shot. Willie toured with the band, but he was gone by the time the final VU record, *Squeeze*, made its way, briefly, onto record-store shelves. It sold even fewer copies than the band's famously low-selling records and is now, of course, a collector's item.

Willie's style was an amalgam of Mick and Iggy and Cher, with a mush-mouthed glam delivery all his own. He was very much a Boston-area mainstay in the late '70s, but outside the region, his two releases with the Boom Boom Band garnered little excitement, favorable press, or sales. Dave Marsh, a prominent rock critic of the era, was not alone in heaping scorn upon *Willie Alexander and the Boom Boom Band* and the follow-up, *Meanwhile...Back in the States*. In the influential and invaluable *Trouser Press* magazine, Willie and the boys were dismissed as a merely competent bar band. But I say they can all take a powder, because Willie's salacious singing style, combined with the ace fretwork of fleet-fingered Billy Loosigian, delivered a fine vintage of '70s rock, unmoored by the desire to say anything too deep.

"Rock & Roll '78" is a wailing tribute to a "local rock 'n' roll band in their rock 'n' roll clothes," powered by Loosigian's siren-like lead lines. Fairly choking out the hard syllables in passages like, "Jungle music in the bunker/Chuck Berry rings a bell," Willie's campy, half-spoken vocal feels like a Quaalude-paced "Walk This Way." The swoon-worthy middle eight is a wordless series of *ooh-wee-oohs* and *woah-oo-woah-oos*, with Willie's signature glottal stops on full display. It alone is worth the price of admission.

I've been curious about Willie Alexander since my pal John Dunton-Downer (a Bostonian, natch) sat me down and played me a bunch of his records one night in the early 1990s. When I reached out to Willie in 2013 via email, to learn about the song's origins, I was thrilled that he responded at all, and especially gratified to receive such a toothsome response.

"I think I originally called it R&R LICK #72 or #73," he wrote. "And it was always about the year it was recorded. But really it was the chord progression that I was putting a number on, because it was one of the two progressions that all the new 1955 music was based on. It was the second progression that the doo-wop groups used, not the fast 1-4-5 say blues form—the other one."

A SONG TITLE	AN ALBUM	A BAND	A LYRIC
"Montana/Autumn/78" – Manic Street Preachers (2009)	*Class of '78* – Buddy Rich (1979)	78 Saab – Canberra, Australia, pop-rock (1990s–2013)	"And the old 78 is skipping" – Elysian Fields, "Bum Raps and Love Taps" (2005)

79

It shouldn't come as a major surprise that the smattering of the world's #79 songs refer to 1979, a year in which new wave was ascendant as the first jolt of punk ebbed. Phil Lynott's "Talk in '79," from the Thin Lizzy frontman's *Solo in Soho*, is a laundry list of U.K. acts that name-checks the bigs (Clash, Elvis Costello) along with thinking punks' favorites like John Cooper Clarke, the Slits, and Nina Hagen.[1]

Tom Robinson gave the world one world-class single in "2-4-6-8 Motorway"—and a numerically minded one at that. But when Robinson looked ahead to the end of the decade in "Winter of '79," he saw an apocalyptic vision ("All the gay geezers got put inside/colored kids was getting crucified"). His assessment may have been a tad extreme, but in a sense he was startlingly prescient. London's early 1979 was marked by social unrest, strikes, and riots, and is now remembered, *pace* Shakespeare, as the "winter of discontent." The government's attempt to deflect criticism led to the infamous headline (and Supertramp album title) "Crisis? What Crisis?" There was even a band called Crisis, whose "U.K. '79" railed against fascism, while in "Back in 79" the Toy Dolls yearn for the carefree, laddish days: "Bobby was a skinhead/but Bobby's hair grew//all for a bird that made him b-b-blue."

"79th and Sunset" is decidedly jaunty for Humble Pie. Singing the praises of an L.A. streetwalker with "nut-crushing boobs and jam tart lips" and resurrecting a line from an old Small Faces song ("So many bad ways to be good"), it's a universe away from the Ramones' "53rd and Third," which has nothing good to say about the life of a prostitute.

Manhattan's M79 bus route begins on the Upper West Side and proceeds eastward across 79th Street, often carrying loads of students from various city schools.[2] Vampire Weekend formed while its members were studying at Columbia University, so they knew the route well. With references to the Khyber Pass, "M79" reads like a plea for multicultural empathy, but the pure ebullience on display trumps any charges of over-cleverness by a long shot. The string section alone is enough to induce bouts of air viola.

THE VERDICT

For their first release on Merge Records, Spider Bags—bumptious garage rockers led by Dan McGee—do a rip-roaring cover of John Wesley Coleman's "Summer of '79." Coleman characterizes his original song as "psychedelic-playful," while the Spider Bags version will curl your hair. I spoke with these self-described cosmic brothers separately in early 2015.

Dan McGee: Being in a punk band in New York, I made a lot of friends all over the country. When I moved down here and started being the Spider Bags, there was this one tour when we were just crisscrossing, and everybody was like, You gotta meet the

---

1 In the Hoodoo Gurus' exhilarating "(Let's All) Turn On," Dave Faulkner rapid-fire name-checks songs and bands of a mostly rarified stripe: "'Shake Some Action,' 'Psychotic Reaction,' 'No Satisfaction,' 'Sky Pilot,' 'Sky Saxon. That's what I like."
2 M79 is also the name of a grenade launcher used in the Vietnam War.

Golden Boys. And the same thing was happening to them. So we just knew, the two bands knew we'd be buddies. Like we were already friends when we met. Wes and I connected first and strongest. Did a lot of touring together. We made a record together in Kentucky, like in a day.

John Wesley Coleman: We don't have to talk too much. When we do, it's funny, but we know when to be serious. I wrote a song about him called "My Friend Dan."

Dan: We had this song we were playing on the road that I was making way more complicated than I had to—I never got a chance to record it—called "Dishrag." Last day of the tour he says, "What's that song? How's it go? I'm gonna record that song when I get back and put it out before you do." And he put it out. He had totally distilled it and made it what I wanted it to be. He called it "Dishtowel." He changed the title! [laughs]. So I wanted to do one of his songs too.

Wes: The song actually had three titles. "Summer of '79" was the first title. I recorded the song for myself at least twice. I cut a version with the Golden Boys, but I changed the name to "Auto Drive." It was a bad idea. Also it's been called "Why Do You Wanna Be a Rolling Stone." The original was recorded in Lincoln, Nebraska. I think it was my first or second 45 as a solo artist and that piqued interest with other labels. People love that song.

Dan: I put "Summer of '79" on one morning, because I was thinking about it a lot, thinking the Golden Boys have so many good songs that the songs get lost on their records. I saw it in my daughter's eyes; she was like 2 years old. I was definitely putting it on to see what it did to her, and she made me play it again. She used to say, "Play it again, Dad. Play it again. Play it again."

Wes: When I moved [to Austin] about 12 years ago, there were these friends of mine and they had a band called The Vain. They looked like Johnny Thunders, they had Keith Richards haircuts. They ended up disintegrating because they were so into creating a myth that they never really took off. I think I was sleeping on the floor of this house, cause I didn't really have a place to stay, and they were there, and we were hanging out a lot playing music. In my head it was like, "Why are you pretending to be this way? You weren't born before '79. You weren't around." I kinda took that sentiment and wrote a riff on that.

Dan: The best verse in Wes's song, it's amazing, it's the one, "Why do you think you're all alone/why do you think your dad is the king of rock 'n' roll?//You weren't there, you weren't alive//go around the world and drive." That's it. You weren't there. You weren't alive. Go see the fuckin' world. And that's what rock 'n' roll tells people: Get in your car and drive.

Wes: Sometimes I write things and I can't even explain any of it, and Dan'll explain it for me.

Dan: People talk about rock n roll being dead, because it was born in this period of American culture where everything was fads. People think that it constantly has to be revolutionary, but changes in art are rarely revolutionary and they happen over a long period of time. Rock 'n' roll was revolutionary. When it came, everybody was like, "It's just like calypso. In two months no one's gonna be buying it." But it's an American art form, like jazz. It doesn't die.

A SONG TITLE	AN ALBUM	A BAND	A LYRIC
"Pieces of 79 and 15" – The Strawbs (1969)	*Live Seventy-Nine* – Hawkwind (1980)	Spot 79 – Portland, Ore., jazz-funk (2000s)	"Wednesday, October 13th, 1982/ Volume 4, Number 79" – The Replacements, "Lovelines" (1983)

No decade in the rock 'n' roll era has been reviled quite like the '80s. While plenty of admirable music was made during that time, collective memory tends to focus on the overbright sheen, gated snare drums, and soulless production aesthetic that dominated the era and characterized its best-known songs. Despite engendering the "college rock" genre, hardcore scenes on both coasts, and a golden age of hip-hop, the '80s will forever be remembered as the time of Phil Collins and Lionel Ritchie, Debbie Gibson and Tiffany and Loverboy, an era when the Stones and David Bowie grew ever more bloated and corporate, and Lou Reed did ads for Honda Scooters.

Imagine for a moment that the laid-back lounge music/Krautrock hybrid of Stereolab emanated not from the mid-'90s but from the mid-1960s, and was vocalized not in earnest yet blasé French but rather in clipped, Teutonic syllables, and you will have a good sense of "Im 80 Stockwerk" by Hildegard Knef. Born in Ulm, Ms. Knef spent time in a POW camp during World War II, became an actress while still a teenager, and performed a brief but controversial nude scene in the 1951 film *The Sinner*. In the early 1960s she switched her focus to singing, releasing hundreds of recordings and selling millions of records. German chanson may be an acquired taste, but "Im 80 Stockwerk," from 1970, is irresistibly cool, delivered with mocking restraint and decked out in the smoky tones of a Germanic Bette Davis. In a piece of 80-centric synchronicity, the second husband of the oft-married Knef was English actor David Cameron, who had a small role in the ghoulish Eric Roberts vehicle, *Star 80*.

You can't make this stuff up.

Eighty comes up in the context of intimate relationships in a pair of songs from the early '80s. "Dancing With My Eighty Wives" is a standout from *Fields*, the lone release by the Jersey-based Individuals. Led by Glen Morrow, a rock journalist and eventual co-founder of the venerable indie label Bar/None Records, the band was in on the ground floor of the brief heyday of the "Hoboken Sound," a loose confederation that included the Feelies, Bongos, and Cucumbers. The desert-evocative "Eighty Wives," which pictures Satan cutting the rug with our hero's harem, transcends its period production and still sounds great. Meanwhile, Kirsty MacColl's steel-drum-spangled "I'm Going Out With an 80-Year-Old Millionaire" reflects her love for Caribbean musical idioms. In typically wry fashion, Kirsty posits the plus side of attachment to an octogenarian as she waits for the old guy to kick the bucket.

THE VERDICT

Echoing the theme of our #70 prizewinner, the topper here is a heartfelt torrent of emotion aimed at a swath of time. But while Mike Watt's "Against the 70s" was written 20 years after the period it lambastes, Killing Joke's "Eighties"—seething, apocalyptic music for an era that valued danceability over soul—is defiantly in the present tense. Over drummer Paul Ferguson's impatient tribal groove, Jaz Coleman unleashes a larynx-shredding vocal: "Eighties—get out of my way/I'm not for sale no more."

The fierce beauty of "Eighties" has been somewhat overshadowed by the controversial coopting of Geordie Walker's central guitar riff by Nirvana, on "Come As You Are." The Nirvana cut not only recycles that lick almost identically, it uses it in precisely the same way: as the song's opener and principal leitmotif.[1] But controversy be damned, "Eighties" is still a sharp jab to the solar plexus of a stiff and corporate time, a black belt fighter undiminished by having performed a lead role—uncredited and uncompensated—in one of the next decade's most identifiable anthems.

A SONG TITLE	AN ALBUM	A BAND	A LYRIC
"80's Life" – The Good, the Bad & the Queen (2007)	*80* – B.B. King (2005)	Tahiti 80 – France, indie pop (2000s–2010s)	"The freezing dark hole/ She's 80 years old" – Tall Dwarfs, "The Slide" (1987)

---

[1] This phenomenon receives book-length treatment in the excellent *Sounds Like Teen Spirit* by Timothy English, whose title derives from that *other* Nirvana song and its considerable debt to Boston's "More Than a Feeling." Further evidence of the prevalence of riff borrowing is the fact that "Eighties" itself bears a striking similarity to the Damned's "Life Goes On," which preceded it by three years.

81

Make no mistake—the number 81 means one thing only: the Hells Angels Motorcycle Club. H and A, the 8th and first letters of the alphabet, have long stood for the notorious 60-year-old organization, or organized crime syndicate, as the U.S. Department of Justice would have it. Fiercely protective of their brand, the Angels have in recent years abandoned the weighted pool cue for the lawsuit, successfully suing numerous infringers, including a clothier whose products were emblazoned with the number. "Eighty-one is Hells Angels," longtime leader Sonny Barger stated plainly to *The New York Times* in 2013.

81: Number of blunts Snoop Lion (nee Dogg) smokes per day

Decades earlier, in 1965, Barger had written a letter to President Lyndon Johnson, offering the Hells Angels' services in the fight against communism in Vietnam. That same year, a dance called the 81 caught on with the youth of Philadelphia. It's not entirely clear what the number refers to, any more than it was clear what the Sharpees' "Do the 45," also released in 1965, was all about. But the kids didn't seem to mind the ambiguity, and when they started doing the 81 to the Martha Reeves & the Vandellas song "In My Lonely Room" at local record hops, it was only natural that a song be written to leverage the dance's popularity. Producer Jerry Ross and his collaborator, Kenny Gamble, responded with a variation on the Martha Reeves song.

They enlisted a trio of Staten Island teenagers—Beryl "Candy" Nelson, her sister Suzanne, and a friend, Jeanette Johnson—known as Candy & the Kisses, to record it on the Philly-based Cameo Records. Cameo's proximity to the studio where Dick Clark's *American Bandstand* was filmed meant that its finger was on the pulse of youth culture, and artists could respond quickly to the latest craze. Gamble and Ross brought in Leon Huff to play keyboard. It was the first collaboration between Gamble and his future partner in the legendary Gamble & Huff songwriting team, which fueled the Philadelphia soul sound of the 1970s.

Featuring the characteristic Motown girl-group bounce, "The 81" opens with a rapidly strummed suspended 9th chord filigree, a proven way of launching a song forward, whether double-time à la Fine Young Cannibals' "Good Thing," or like syrup, à la Prince's "Kiss." It then kicks into a mostly tambourine-and-handclap beat, with a vocal by Candy Nelson that's like a cub to Martha Reeves's lion. Despite its simple charms, the record never really caught on, and follow-ups fared even worse. A new name (Honey Love & the Love Notes), a label switch, and new songs penned by the young team of Ashford and Simpson could not reverse the group's fortunes. In 1969 Candy & the Kisses hung it up for good. Forty years later, Jeanette Johnson wrote a blog post expressing appreciation to those who have kept the flame alive: "You just can't help dancing whenever "The 81" comes on the radio," she wrote, "because it was a really excellent jam."

Joanna Newsom has never written a song that could be properly called a jam. Her compositions are far too high-minded for that label, and as with many purveyors of serious, arty music, no one is neutral about this singer/composer/harpist. In fact, her ability to provoke extreme reactions is impressive. Even listeners whose iPod playlists include so-called freak folk artists like Devendra Banhart might draw the line at Newsom, whose idiosyncratic, childlike vocals can resemble the sound of an Our Gang member with poetic aspirations. As an extreme and not particularly universal example, my own mother, struggling to remember the name of the singer she'd recently heard on NPR, was barely able to form words—so rich was her disgust—until she enunciated the succinct-enough fragment "the girl with the harp."

"'81," the third track from Newsom's sprawling third release, *Have One On Me*, is one of her most accessible songs—a mother could easily love it, just not my mother. "Meet me in the Garden of Eden," she sings with disarming directness. "Bring a friend/ we are gonna have ourselves a time." The song is a world unto itself, like being in a glade, and anyone looking to see what the fuss is all about would do well to start here.

THE VERDICT

Deerhoof's "+81" begins with a few bars of brass fanfare followed by a burst of rolling, marching-band percussion and an exploded Chuck Berry lick. Enter Satomi Matsuzaki, her slightly flat, guileless vocals riding an insistent insistent 4/4 kick-drum pulse. The song's opening line ("The building building from the side to side") is also strangely Chuck Berry-esque, echoing the "bumper to bumper, rollin' side to side" bit in "Maybellene," along with any number of Chuck's songs that lean hard on one note, serving a rhythmic function as much as a melodic one. It isn't until the chorus rolls around—"Choo choo choo choo beep beep!"—that "+81" launches itself into a realm far beyond anything Chuck ever imagined. Into another dimension, or at least another key. Rather than a functional middle eight, the melody meanders in little spirals, like a girl singing to herself while playing jacks. It's no small feat that the song finds its way back to its weirdly insinuating vocal hook. Like that painting of the Virgin Mary that incorporated elephant dung, "+81" skews an accepted form via the use of unexpected materials.

A SONG TITLE	AN ALBUM	A BAND	A LYRIC
"Ward 81" – The Fuzztones (1980)	*The '81 Demos* – Weekend (2014)	Bumblebeez 81 – New South Wales, Australia, indie rock (2000s–2010s)	"He was a diplomat's son/ It was '81" – Vampire Weekend, "Diplomat's Son" (2010)

82

Kurt Vonnegut's breakthrough novel, *Cat's Cradle*, hinged on a deadly substance called ice-nine. On page 1 of an earlier novel, *The Sirens of Titan*, he posits a world in which humankind has overcome unhappiness by figuring out that there are 53 portals to the soul. In *Breakfast of Champions*, Vonnegut's oft-used character/doppelganger, Kilgore Trout, is reported dead after publishing "his two-hundred-and-ninth novel." In *Happy Birthday Wanda June*, when the character Harold is asked whether he's killed women, he responds "seventeen of them—eleven by accident." And so on.

Clearly, the author of *Slaughterhouse-Five* had a thing for numbers—the finiteness of them, their unique descriptive properties. I tell you all this because Vonnegut's *Hocus Pocus*—published in 1990, long after his peak of popularity—tells the story of Eugene Debs Hartke, a Vietnam veteran-turned-professor who loses his job due to outspokenness and ends up working at a prison. Hartke becomes obsessed with calculating two quantities: the number of women he's had sex with, and the number of people he killed during the war. He ends up finding that he has killed and made love in equal measure: 82 times each.[1]

Once, when Vonnegut was moved to squeeze his most heartfelt beliefs about creative writing into eight laws, he praised Flannery O'Connor for breaking "practically every one of my rules but the first."[2] Not often does one get to refer to Bruce Springsteen and Vonnegut in the same breath, but The Boss shares a love for O'Connor to match The Snarf (Vonnegut's high school nickname: I kid you not). "There was some dark thing," Springsteen said, "a component of spirituality—that I sensed in her stories and that set me off exploring characters of my own."

Before he got down to exploring characters, Bruce tended to simply people his verses with them and give them colorful names. In what has to be the briefest of all Springsteen songs—"Does This Bus Stop at 82nd Street?" (a lean 2:05)—he packs in a parade. It's not quite as memorable as the cream of *Greetings From Asbury Park*, but the song is a fine collision of "Stuck Inside of Memphis"-style Dylan and Van Morrison's urban phantasmagoria, transferred to the Boardwalk. And *you* try writing a line like, "Wizard imps and sweat sock pimps/Interstellar mongrel nymphs." It's harder than it looks.

THE VERDICT

The Shins' "Fall of '82" is a personal song. It makes no reference to any of the key events of the season in question, which included a deadly earthquake in Yemen, the dedication of the Vietnam Veterans Memorial in Washington, D.C., and the first gig by the

---

1 *Breakfast of Champions* also contains a reference to an errant *New York Post* article that lists Kilgore Trout's name as Kilmer Trotter and his age as 82. That book is fairly loaded with numerical references: Several male characters are introduced by name and the dimensions of their sex organs, e.g., "Robert Petko…was a career officer in the Army. He had a penis six and one-half inches long."
2 Rule #1 is essentially "Don't waste the reader's time."

Smiths, at the Ritz in Manchester. Shins main man James Mercer remembers it as a time when his older sister reappeared at just the right moment to snap him out of an advanced case of adolescent angst. The song showcases Mercer's deftness at hitting the sweet spot without losing complexity. Somewhat at odds with its title, the song's sonic touchstones derive from the previous decade, when rock radio melodies were straightforward in the manner of Joe Walsh's "Life's Been Good," the verses of which it resembles not a little. But rather than being derivative, "Fall of '82" feels like a brave revisitation of the era. After a blissfully creamy horn break, the action melts into an unexpected interlude that veers away from formula into fascinating.

A SONG TITLE	AN ALBUM	A BAND	A LYRIC
"82 Afros" – Camp Lo (feat. Ski) (2007)	*82* – The Fevers (2005)	82 – Kenya, electronic (2000s)	"His eighty-two brings many fears" – Keith West, "Excerpt From a Teenage Opera" (1967)

83

Eighty-three may be the 23rd prime, but—23 Enigma proponents be damned—it's hard to find a universal component to 83. It's a sacred song (Psalm 83 exhorts God to smite the enemies of Israel); the atomic number of bismuth (abbreviated Bi, thus 83 *might* be slang for bisexual among science nerds); and it's a highway that passes through Cupertino, Calif., home of Apple. But not a whole lot else.

As for #83 songs, most fail to represent the artists at their best. Ian Hunter's "Apathy 83," for example, has the urban fever-dream imagery of early Springsteen. And while it deftly distills the corporate music-biz ethos of 1976 ("There ain't no rock 'n' roll no more/just the music of the rich"), "Apathy 83" feels a little tepid next to Hunter's A-list material. In "Moon 83" the B-52's reworked the superior "There's a Moon in the Sky Called the Moon" from their audacious 1979 debut.

Far better is Mission of Burma's swirling "SSL 83," the ostensible title track from *The Sound The Speed The Light*—their third LP after reuniting in the early '00s. Mission of Burma's initial run was brief, its dissolution in 1983 hastened by guitarist Roger Miller's tinnitus. That same year, Ian Astbury formed the Cult (from the ashes of "Southern Death Cult") with guitarist Billy Duffy. "83rd Dream" is a standout from the band's debut a year later, a powerful slow burner that showcases Astbury's histrionic wail and Duffy's spidery arpeggios. The song shares some similarities with the winner here. Both songs have "dream" in their title, and both bands draw on the legacy of the Doors. In the end, however, "83rd Dream" is no match for this category's top dog.

THE VERDICT

It's almost a shame to use "New Gold Dream (81, 82, 83, 84)" solely to fill the #83 slot. The song is big enough, numerically minded enough, and timeless enough to occupy this entire barren stretch of land, yet it would be a disservice to Candy & the Kisses and Deerhoof and the Shins to cede a four-digit swath to Simple Minds. But I won't say I wasn't tempted.

New Gold Dream, Simple Minds' breakthrough LP, had chart hits in the slick '80s funk of "Promised You a Miracle" and the stately "Glittering Prize," yet the title track best captures the band's grandeur and measures up to its arena-rock aspirations. Arriving on a lolloping synth pulse, the song breaks into a monster big-beat groove that manages to stay supple, with a prominent on-the-one cowbell and a dreamily self-assured vocal from Jim Kerr. And that chorus is a marvel, managing to imbue a succession of crooked numbers with a sense of pure triumph.[1]

---

1 "89-90-91-92" lacks the zing of the Simple Minds classic, but then, it was never meant to be sung, not least by Jim Kerr. This miniature by the minimalist composer and John Cage acoyte Michael Nyman consists of four sections of increasing length, all based on a theme from a Mozart symphony.

If 6 Was 9

A SONG TITLE	AN ALBUM	A BAND	A LYRIC
"83"– John Mayer (2001)	*The Original Sound of Sheffield, '83–'87* – Cabaret Voltaire (2001)	M83 – France, electronic/ dance (2000s–2010s)	"Will you live to 83?" – R.E.M., "E-Bow the Letter" (1996)

LAST WORDS:
"I shot the happiest 83 of my life."
– Chi-Chi Rodriguez, on drinking a bottle of rum to stave off nervousness at his first Masters tournament.

A few theories exist as to why the small rural town of Smithtown, Pennsylvania, came to be rechristened Eighty-Four. Some believe the name change (which placed it in the rarified company of Ninety Six, South Carolina, Eighty Eight, Kentucky, and other "Places Named After Numbers," in the words of Frank Black) commemorated the 1884 inauguration of Grover Cleveland. Others believe the renaming came down to the whim of a particularly uninspired postal clerk.[1]

Almost exactly one century later in Edinburgh, Scotland, teenage Marcus Eoin Sandison joined his older brother Mike's band, which they called Boards of Canada. Eventually, in the course of a few records, the brothers created a style of trippy electronic music that has been revered, often imitated, but never quite equaled. Mike Sandison gave perhaps the perfect description of the duo's sound when he told Pitchfork that they loved "the idea of making music where it's really difficult to figure out which instruments you are listening to but you just don't care." "84 Pontiac Dream," from 2005's *The Campfire Headphase,* wafts across the room in an enticingly degraded mist, a dream about a dream of a car, with a shard taken from an '80s corporate training video providing a touch of faux futurism.

THE VERDICT

Music experienced a sea change in the late '60s, from a melodic base to songs based on insistent, visceral riffs. "Doom 84," by Screaming Females, is built around a diabolically fierce lick that combines the collective thrills of *Fun House*-era Stooges, dark slabs of Sabbath's Tony Iommi, and the molten-yet-precision-cut shredding of Hendrix. It's something like "When the Levee Breaks" played at a gallop, fantastically heavy but never lumbering.

Perhaps my upbringing has caused me to associate this degree of musical heft with testosterone, but the New Jersey trio led by the diminutive, lion-voiced Marissa Paternoster demanded a reconfiguration of my retrograde notions of hard rocking. "Doom 84" is simply seven minutes of catharsis. Paternoster's voice, with its wide vibrato suggestive of sheer abandon, comes straight from the gut. And the blunt-force impact of her voice does not stop at her lyrics, which have an audacity to match the guitar pyrotechnics and demonic howl: "You're piss on my pillow/you're filth in my veins/and I will make sure that I don't make mistakes."

One could be bowled over enough by "Doom 84" to not even give a passing thought as to the number in the title, which is nowhere to be found in the song. Eventually I began to wonder, though, and when I couldn't find any specifics online, I contacted Marissa Paternoster in early 2013. She provided me with the answer to my question as well as crucial insight into her creative process.

---

1 Thanks to the Avalon Motor Inn, a local motel, Eighty-Four, Pa., has been immortalized on the silver screen, in *The Mothman Prophesies.* Mark Pellington, who directed the film, also directed the haunting video for "74-75."

"I usually don't talk too much about the lyrical content of our songs 'cause a lot of it is kind of abstract, and to be quite honest, sometimes I find it difficult to draw meaning from my gibberish. I often work with lyrics from more of a phonetic perspective—I choose words that sound "correct" and then draw out meaning after the fact."

Her response here reflects one of the basic truths of rock: lyrics don't always mean that much, and they don't have to—as long as they sound right within the song. I wish more bands and writers would admit that.

"When we first began to put "Doom 84" together, we simply called it "Doom" for reference. When Jarrett [Dougherty, Screaming Females drummer] and I were tracking some drums for a demo, the metronome was set to 84 beats per minute, so we called the song "Doom 84." Luckily, the title sat nicely with the pace and tone of the song. It stuck."

A SONG TITLE	AN ALBUM	A BAND	A LYRIC
"84" – The Fucking Champs (1997)	*EB 84* – The Everly Brothers (1984)	Combat 84 – England, oi punk (1980s)	"Well she's really going fine for eighty-four" – Elton John, "Country Comfort" (1970)

85

I was all set to disqualify "Nineteen Hundred and Eighty Five," the final track on Paul McCartney & Wings' *Band on the Run*, on technical grounds (see Ch. 19 for a discussion of the so-called Prince Conundrum) when I realized that the number in question actually *does* stand alone. Now, it could well be argued that "Nineteen Hundred and Eighty Five" is the best-known instance of 85 in pop music. *Band on the Run* sold a zillion copies, and Macca has played it as recently as the 12-12-12 Concert for Sandy Relief. There's just one problem: lyrically it is one of the weakest things Paul—*the most successful recording artist in history*—has ever done. The sticking point is not that the words are trite or obvious or uninspired, which they are. That's not a deal-breaker (I have no problem with "Hello Goodbye," for example). It's that they portend something that they don't deliver, a failing I have taken Chuck Berry and Paul Simon to task for, and I believe Sir Paul must be held to the same standard.

Here's what I mean. The extremely catchy syncopated piano figure that anchors the song, along with the opening couplet, sets the scene for a classic sci-fi future epic along the lines of Bowie's "1984," or even Zager & Evans's "In the Year 2525." But once he gets past the opening line ("No one left alive/in nineteen hundred and eighty-five will ever do") McCartney completely abandons any pretense of following through on the initial premise. Instead, it's all about his inability to get enough of "that sweet stuff [his] little lady gets behind." The words are dummy text that Paul never got around to changing into proper lyrics. No wonder the working title was "Piano Thing."

Paul might not have given a toss for 1985,[1] but several acts found the year inspiring. In Snoop Dogg's "Take It Back to 85," the rapper sometimes known as Snoop Lion recalls an era of "Wallabees and Crocasacks/Relics, gun pellets and blue golf hats." "1985," a hot-selling single in 2004 by Gym Class Heroes, looks back at the year Reagan was reelected and recalls "watchin' Smurfs, eatin' Cinnamon Life." Rilo Kiley, the indie pop outfit led by the now-solo Jenny Lewis, came to prominence on the soundtrack of a Christina Ricci movie with "85," a stirring ode to the lingering fallout of a failed relationship. Heck, even the Bay City Rollers weighed in on the subject, in "85," a song that wonders, "Where will you be in '85?" The answer, for the Rollers anyway, was "kaput," putting them directly at odds with the intoxicating Broken Social Scene instrumental "Alive in 85."

THE VERDICT

"85 Weeks" exemplifies the more refined side of Frank Black, who found a more conventional, even mature songwriting style after the dissolution of his famous first band, Pixies. The song is the restrained moment on *Pistolero*, his second record with the Catholics:

---

[1] Indeed, it was not exactly a banner year for the world's foremost left-handed bass player. His biggest moment of the year was his performance at Live Aid, which was pretty much sunk by technical difficulties.

If 6 Was 9

some acoustic strumming, minimal bass and drums, and Black's intimate, decidedly unhowled vocal. While not a straightforward narrative, the song is based on an anecdote told by Black's collaborator, former Pere Ubu keyboardist Eric Drew Feldman, about the chronic awakeness that plagued the great rock iconoclast Captain Beefheart. According to Feldman, his longest stretch without proper sleep was 85 weeks.

The chord progression is deeply Beatle-esque in the "Cry Baby Cry" vein, with a refrain that revolves around Black's made-up locution "un-somnambulist." True to form, he does not completely eschew surrealism: "The world to him was all covered in fur," he sings of the napless Captain, employing an image that smacks of Black's longtime muse, Salvador Dali.

A SONG TITLE	AN ALBUM	A BAND	A LYRIC
"85 on 85" – 6 String Drag (1997)	*Selected Ambient Works 85–92* – Aphex Twin (1992)	85 Decibel Monks – Iowa City, hip-hop (2000s–2010s)	"By the time we got into Tulsatown we had eighty-five trucks in all" — C.W. McCall, "Convoy" (1975)

86

When the English rock mag *NME* issued an unprepossessing cassette compilation called C86 with its May 1986 issue, the odds were low that it would set off a movement. The assembled bands, culled from England's vibrant independent music scene, were fuzzy and strummy and shambolic, delivering simple chord patterns with more passion than precision. Just as punkers reclaimed the rock music they felt excluded from, so the C86 bands reacted against an era of slick dance beats, macho guitar theatrics, and style over substance. True to the punk ideal, the C86 bands made a virtue of amateurism, but—and this is crucial—male-centrism and aggression were 86'd. Another hallmark of the C86 bands was that singles were issued with primitive record-sleeve art, in the 7-inch (not 12-inch) singles format—a conscious rejection of current music-biz norms.

The C86 moment was brief, but what a moment: Small labels popped up like mushrooms to release music by bands formed weeks earlier, by people who'd only recently learned to play instruments. When it was over, most of the bands disappeared or dispersed. But for a while, the C86 effect was seismic. It would be an overstatement to say the twee C86 bands made the world safe for the guilelessness of vocalists like Ben Gibbard, but there, I've said it. *Transatlanticism* by Death Cab For Cutie, the band Gibbard leads, may not have had a seismic effect, but it earned almost universal acclaim upon release in 2003. With its shimmering acoustic guitar filigrees and Gibbard's choirboy vocals, "Expo 86" is one of the record's most spritely and unfettered moments.

Ben Gibbard was born in 1976, the same year country/pop singer Narvel Felts had his final Top 10 hit. Like Gibbard, the Arkansan Felts had a powerful tenor, but Felts' secret weapon was a falsetto that put him within yodeling distance of Roy Orbison and Gene Pitney. But unlike those clarion crooners, not to mention Ben Gibbard, Narvel the Marvel waited a long time for real chart success. His "86 Miles"—a prototypical "I'm *this many* miles from my gal" song—came out in 1967, about halfway between his biggest hit (1975's "Reconsider Me") and the beginning of his recording career proper in 1959.

Wondrously enough, 1959 also saw the first official instance of the verb "eighty-six," (or "86"), which had been incubating in restaurant parlance for decades. Originally the term meant an establishment's refusal to grant admission to an unruly patron ("His antics got him eighty-sixed from the Brown Derby"), but its meaning has grown more expansive, referring these days to exclusion in general, as in "86 the Mayo," by New York ska revivalists the Slackers.

Clearly, a song called "The Mystical Path of the Number 86" is manna to certain kinds of people, including the type who would write a book like this. This nine-minute sprawl of craftily woven feedback noise, from a 1999 LP of the same name by Davis Redford Triad, sounds as if it could have emanated from a stack of Marshalls with a dozen guitars plugged in, manipulated like a giant Theremin by an unseen hand. The man behind the project is Steve Lobdell, a prolific, self-taught guitarist from Portland, Oregon, whose music combines experimentalism, noise, and psychedelia.

Lobdell wrote and recorded "The Mystical Path" in 30 hours, in a barn. He spoke with Portland zine *Bananafish* in 1999, detailing his mental breakdown and weeks spent in a mental institution. The song, he explained, has to do with "the mystical path of banishment, like when you're 86'd from a bar." [Aha!] But more generally, he said, "'The Mystical Path of the Number 86' is all about the cycle of going crazy." Indeed the chaotic squalls and caustic clouds of pulsing noise take Syd Barrett's excursions even deeper into pure chaos; and unlike early Floyd freak-outs like "Interstellar Overdrive," there's no return to relatively safe ground here. So what drives Lobdell? "I don't like overusing the word 'spiritual,'" he said, "but to me music is a spiritual art form."

Billie Joe Armstrong of Green Day[1] expressed a similar sentiment to an interviewer for *Out* magazine in 2010, stating that punk rock is "kind of spiritual in its own way, because people fight over it so much and the meaning of it." He also said punk rock was "all about community," which has at times been a sore subject for him and his mates. After paying dues in the rough-and-tumble Northern California punk underground in the early '90s, Green Day reached the major leagues in 1994 with *Dookie*, which went on to sell 15 million copies. In the process, a few bridges were burned. On "86," Armstrong recounts the experience of being shunned—86'd as you might say—by the Berkeley punk mainstay known as Gilman as well as by former fans because he and the boys had gone out and gotten famous. Armstrong makes no apologies. The refrain ("There's no return from 86/Don't even try") confirms that the line has been crossed.

THE VERDICT

Yo La Tengo's "86-Second Blowout" moves at a gallop, squeezing in three vaguely dire verses but never returning to an image or a phrase. No wonder the song ended up being named after the fact, by random factors. I had always taken it for granted that Ira Kaplan had titled "86 Second Blowout" because it lasts 86 seconds and it's a blowout.[2] But Jesse Jarnow, author of the incisive YLT biography *Big Day Coming*, hipped me to another, more esoteric and cooler meaning. In Jarnow's reckoning, Kaplan, noting the length of the as-yet-nameless track he had just recorded for *May I Sing With Me*, remembered "96-Second Blowout," a B-side by a pre-dB's Peter Holsapple. To Ira Kaplan, who tends to go the impressionistic route when naming songs, this was pure manna. "86-Second Blowout" seems sonically indebted to the Feelies, who had a thing for knotty, strummed patterns like this one. With Georgia Hubley's nimble fills whipping the song along like a crop-wielding jockey, this sprint is bracing from wire to wire.

A SONG TITLE	AN ALBUM	A BAND	A LYRIC
"86" – Bonnie "Prince" Billy (2000)	*Expo 86* – Wolf Parade (2010)	Agent 86 – Northern, California, punk (1980s)	And if it wasn't for the '86 World Cup/ it wouldn't have got this far" – McLusky, "Dethink to Survive" (2002)

---

1 Green Day's songbook is notably rich in numerical bounty, including "Sweet 16," "21 Guns," "80," and "99 Revolutions."
2 The song lasts that long anyway. The MP3 version rather disappointingly clocks in at 1:32 due to a brief puff of silence at the end.

In a classic Bob Newhart routine, Lincoln's press agent learns that the president has made "a couple of minor changes" to the Gettysburg Address, including revising "four score and seven years ago" back to 87. The flak advises him to leave it alone: "Abe, that's meant to be a grabber. We test-marketed that in Erie and they went out of their minds." Indeed, by scrawling "Four score and seven years ago" on the back of a mythical envelope in 1863, Abraham Lincoln achieved oratory immortality with his conspicuous avoidance of the word *eighty-seven*. And people have been avoiding it ever since.

In a spot like this, one might be grateful for the existence of a bona fide #87 song by a bona fide star. But there's not much to say about David Bowie's "'87 and Cry." It comes from an album that Bowie himself called his nadir. Now, it's always wise to take a musician's perspective on his own career with a grain of salt, but in the final analysis, Bowie's own damning assessment of 1987's *Never Let Me Down* holds up. It wasn't the record's commercial failure that bothered him—this happens in art—what he hated was that there was no inspiration behind the project. It was not "heartfelt." As with any great musical failure, the record flirts with ludicrousness, like the Jack Black-worthy narration on "Glass Spider." That's the thing about nadirs: they're a lot of fun to look back on.

So it's unfortunate that "'87 and Cry" is not worse. It's still pretty god-awful—processed hard rock with aggressively huge drums, not a hair out of place—but on a record that includes Mickey Rourke "rapping," the song lands somewhere in the middle, like the lukewarm water of Derek Smalls. It's telling that in my informal survey of reviews, not a single critic said a bloody thing about "'87 and Cry" in the course of slamming the record. Perhaps most revealing is the song's ignominious placement in the track list: second-to-last, "the devil's spot," wherein an artist, all but admitting the existence of space to fill, inserts something less than crucial before issuing a final big bang. In the case of *Never Let Me Down*, the final whimper is a cover of Iggy Pop's "Bang Bang."

THE VERDICT

In spring 1987, a few months after David Bowie hit bottom, New Order reached an early career peak with *Substance*. This collection of the 12-inch versions of all their singles up to that point also included a new track: the rapturous "True Faith," which might have waved hello to the title track of *Never Let Me Down* while sailing into the Top 10 a few months later.

As with many New Order tracks, there's not even a mention of the title in the song, much less an 87. But what of '87? Usually this naming trope applies to remakes of old songs, à la "La Bamba '87," "When a Man Loves a Woman '87," and "Glad to Be Gay '87." New Order's "True Faith" was released in '87, so what would be the point of adding the year? *Substance* was sometimes dubbed *Substance 87* to distinguish it from the concurrent Joy Division collection of the same name. But "True Faith" became "True Faith

'87" in the digital age, presumably to distinguish it from "True Faith '94" and "True Faith – 2011 Total Version" and the "True Faith (Shep Pettibone Remix)."

It really doesn't matter. "True Faith"—whether '87 or by any other name—is a landmark New Order single, right up there with "Blue Monday," uniting a sort of German automotive precision with the spine-loosening dance groove and Bernard Sumner's sugary yet tart vocals. Yes, the title's a stretch, a record company contrivance, but that *is* what it's called, and you can buy it for a dollar.

Good enough for me.

A SONG TITLE	AN ALBUM	A BAND	A LYRIC
"Pontiac '87" – Protomartyr (2015)	*87* – Mercedes Sosa (1993)	Group 87 – L.A., jazz fusion (1980)	"I've been working on this palm tree for eighty-seven years" – Neil Young, "Last Trip to Tulsa" (1968)

88

"Rocket 88," a hit for Jackie Brenston & His Delta Cats in 1951, has earned a reputation as the first true rock 'n' roll song. Consider that for a second: the first rock song ever, and a number song at that. Not everyone agrees—in fact, in Nick Tosches's invaluable *Unsung Heroes of Rock 'n' Roll*, in the chapter devoted to Brenston and his all-too-brief moment in the sun, Tosches lays waste to the claim of "Rocket 88" as the ur-rock song. He points out the absurdity of trying to pinpoint the first of *any* sprawling art form, e.g., the modern novel, and makes a compelling case that the claim was the work of the self-aggrandizing Sun Records honcho Sam Phillips.[1] Nevertheless, the claim has been repeated so often that it has become part of the story. (For example, on YouTube, an audio-only version of "Rocket 88" includes the description, "Widely acknowledged as the first 'rock and roll' song.")

There's also the controversy over who gets the credit for writing the allegedly first rock song ever. Five songs came out of the session that yielded "Rocket 88," four of which were divided into two singles. The single credited to the band's leader, piano player Ike Turner, along with His Kings of Rhythm, bubbled under; Brenston's A-side achieved immortality. It comes as no surprise that this situation did not please Ike Turner in the least.

What no one disputes is the greatness of "Rocket 88." The song was a major signifier in the development of rock, a new amalgam that incorporated root elements of country and R&B with wailing sax, distorted guitar (apparently by lucky accident), and Ike Turner's "careening glissandi and manic triplets," in the words of Tosches, all decked out in lurid, deep-in-the-red overdrive. Brenston's lusciously soulful vocal tops the proceedings, his love of boozin' and cruisin' fully palpable.

Just as Jackie Brenston will always be known for a song about a proto-muscle car with a Hydra-Matic drive,[2] the Nails, a New York-via-Boulder, Colo., band, are best known for the 1981 single "88 Lines About 44 Women," which has automotive aspects of its own.[3] Starting with Deborah ("a Catholic girl, she held out to the bitter end") and finishing with Amaranta ("Here's a kiss/I chose you to end this list"), "88 Lines" plays like a first cousin of Jim Carroll's "People Who Died." Its eerie minimalism, and Marc Campbell's deadpan litany, has a strange and enduring allure. In recounting the ones that got away and those he was lucky to leave behind, the song is a marvel of obsessive determination, fulfilling its title premise with wit and precision.

---

[1] Tosches argues convincingly that if a date could be slapped on when rock began, it would be closer to 1946–47, with the proliferation of songs like Wynonie Harris's "Good Rockin' Tonight" and the emergence of blues/country amalgam dubbed "hillbilly boogie." Nevertheless, crucial elements, such as the terminology and nakedly lusty sentiment of songs like Trixie Smith's recording of "My Daddy Rocks Me (with One Steady Roll)" in 1922, had been around for decades. Accompanying a YouTube recording of a 1938 version of the song are numerous posts attesting to *its* status as the very first rock song. On his 1956 A-side "Old Time Rock and Roll," New Orleans-born R&B man Champion Jack Dupree declared, "We've been doin' this since 1929/ but the disc jockeys and the teenagers just heard it."
[2] "She Took My Oldsmobile" by the Romancers—East L.A. Chicano garage-rockers circa 1966—is punctuated by anguished cries of "My 88, My 88!"
[3] E.g., the line "Dinah drove her Chevrolet into the San Francisco Bay," and the song's appearance in an award-winning TV ad for Mazda.

255:
Number of seconds in "88 Seconds in Greensboro"

Referring to a piano as the 88 for its 88 keys is a legendary locution whose echoing vowels tinkle onomatopoetically. "Lemme hear a little of that bass … with those 88s," is the ecstatic exhortation in the Capitols' hit from 1966, "Cool Jerk."

In 1983, a year after the Go-Go's recorded their own version of "Cool Jerk," PBS aired "88 Seconds in Greensboro," a documentary that reconstructed the fatal shooting of five demonstrators by members of the KKK and the American Nazi Party. Upon viewing the doc, Andy McLuskey of Orchestral Manoeuvres in the Dark was moved to express his outrage.

To be sure, OMD were more interested in man's inhumanity to man than your average synth-pop band, as their early hit "Enola Gay" shows. Unlike "Enola," though, "88 Seconds in Greensboro" misses the target. Not knowing a thing about the real Enola Gay, one can feel the foreboding in that tremulous keyboard figure, like the path of the plane on its inevitable course. But somehow McClusky's urgent repetition of the phrase "88 seconds is all it takes" over an unwavering three-chord progression fails to leverage the drama of the event that inspired the song. Reduced to a celebratory terrace chant, it feels at odds with itself.

THE VERDICT

It has been clear from the start that #88 would involve major soul-searching. On the one hand, "Rocket 88" is a foundational song—whether it's the first rock single, or just one of the first. The story behind the song has controversy in the dispute over authorship, and tragedy in the sad demise of Jackie Brenston, who, after failing to capitalize on his hit, eventually had to humble himself and return to Turner and his Kings of Rhythm as a side sax player, and wound up boozing himself into an early grave.

While fully aware of the importance of "Rocket 88," I cannot get past the fact that it was "88" by Anna Domino that started me on number songs in the first place. What's more, it was a very specific kind of enthusiasm for a song that pulses below the radar that inspired this quest. Unlike "Rocket 88," the story of Anna's "88" has never been told. Even the story of Domino herself has only been told in broad strokes.

She was born Anna Virginia Taylor in Tokyo, an Army brat, and she'd continued to hop the globe while pursuing a musical career, scoring a record deal with a French label and operating out of Brussels, Paris, and other far-flung places. I'd first heard her as a collaborator with the 6ths, Stephin Merritt's numerically named side project. After encountering the epic "88" on that now-legendary mixtape, I combed through reviews and interviews, looking for some insight into the song, but turned up nothing. Happily, I eventually discovered the existence of a butterfly known as Anna's Eighty-eight (*Diaethria anna*), a black, white, and red arthropod that looks as if its wings have been graffiti-tagged "88" in bubble letters. I'd paid attention to vague signs before, but this one couldn't have been any clearer: Anna + 88 + Made by God. Who could argue with that?

So I wrote to Anna Domino, telling her about this book and the place her song had in its genesis. I also told her about my current dilemma, and asked if she could tell me more about the song and its inspiration. A few hours later, she responded.

First off, yes, the title is a reference to the year in which the song was written. "1988 was an ecstatic and horrifying year in my life," she wrote. "Great highs (love of my life, lots of travel and sleepless, inspired nights) and deep sorrows (loss of great friend and first love, poverty, madness and hunger too close)."[4]

Her explanation of the song went deeper:

"Writing music, for me, is always about describing a place, the same places over and over, the same emotional atmospheres in landscapes—you do get closer to that. I always told people that "88" was about doing drugs because that would get a laugh out of them and answer the question. It isn't though…Hard to describe (hence above) but I will try. At the far reaches of the imagination there is a place of such exalted beauty that to glimpse it is to never forget it. It is where great art comes from. The pursuit of it is what drives the search for god and can easily cause a break with reality and a slip into madness. Fine line there.

"Some artists and divines have lived in its presence (William Blake, pick a saint) but not many people can stand sustained exposure. I have felt/seen it only twice. It did not come as a vision exactly but the lyrics to the song were written as the brief, violently intense insight faded. A great place to visit, to know is there, to hope is where we go in death, but I wouldn't want to live there. Things have cooled considerably since, farther from God, but I got to live."

A SONG TITLE	AN ALBUM	A BAND	A LYRIC
"Delta 88" – X (1980)	*88* – Fats Domino (1989)	88 Fingers Louie – Chicago, punk (1993–2010s)	"Well I gonna pick you up in my 88/ Get ready sugar now don't be late" – Jerry Lee Lewis, "Wild One"

---

[4] As for the "Rocket 88" vs. "88" conundrum, Anna sympathized with me: "'Rocket 88' does represent a powerful moment in pop music history," she wrote, "a really tough choice. They're completely different worlds in a way. I learned to drive in that car."

Is bird imagery ever a positive sign? R.E.M.'s *Document*, the band's final record for the indie label IRS, was filled with it, and though the LP contained what may be the band's best known song "(It's the End of the World as We Know It)," some of the R.E.M. faithful thought its big-rock production and tendency toward scrutability was strictly for the birds.

Birds were nowhere in sight on the follow-up, *Green*, although a chorus of crickets heralds the hymnlike "You Are the Everything." It was a bigger, bolder, more camera-ready R.E.M. that emerged on rockers like "Orange Crush," which found Peter Buck swapping out his ringing Rickenbackers for a room-filling wall of kablam. Divided between the Air Side and Metal Side, with an inside sleeve that provided lyrics to *a single* song, *Green* was self-consciously arty. You can't blame them for wanting it both ways: the clear enunciation and straightforward rock sound, but also the mystique. *Green*, right down to its orange cover, was R.E.M. not wanting to be pinned down. Yet the band that gave us "1,000,000," "9-9," "Driver 8," and "Seven Chinese Brothers" didn't begrudge its audience a couple of number songs. The lovely, lilting hidden track, "11," helped set off a trend for secreting songs within the generous playing times of CDs, and the opener is an absolute killer.

THE VERDICT

"Pop Song 89," all rhythmic tumult and spiky guitar riff, serves as something like a reintroduction to what we would now call R.E.M. 2.0: "Hello, how are you, I know you I knew you, I think I can remember your name." The implication is that Michael Stipe's just mocking the jaded rock star pose as he tries it on for size, but clearly he kind of likes it. In any case, what a great way to start a song, or an album—just say hello.[1] Cadence-wise, the clearest hello descendent of "Pop Song 89" is the Doors' "Hello I Love You"—it's in that little pause right after the greeting. But Stipe's hurried delivery more recalls Jagger's "Hi, hello, how are you/I guess I'm doing fine" in "Get Off of My Cloud." The crucial difference is that the Stones are trying to duck those who would crash their cloud, while Stipe is out there among the crashers, being semi-available and commenting on the vapidity of stardom.

A SONG TITLE	AN ALBUM	A BAND	A LYRIC
"D Song '89" – The House of Love (1990)	*At 89* – Pete Seeger (2008)	The '89 Cubs – Omaha, Nebraska, indie rock (2004)	"Tourists, 89 flowers on their back/ inventors of the Accu-Jack"– Prince, "Sexuality" (1981)

---

1 Others of this sort include "Hello Stranger" by Barbara Lewis, "Hello Again" by The Cars, "Hello Hooray" by Alice Cooper, and ELO's "Telephone Line." In "Smells Like Teen Spirit," Kurt Cobain lingers menacingly on "Hello, hello, hello, hello, how low?" while Gary Glitter's "Hello Hello I'm Back Again" was a favorite of Oasis, who nicked it for their own song titled simply "Hello." Wire Train—dreamy guitar poppers of the early '80s—struck hello gold with "Chamber of Hellos." And, as much as I try to forget, there's that Lionel Ritchie monstrosity.

90

In 1972, sci-fi writer Theodore Sturgeon expressed the idea that "90 percent of everything is crap." Eventually this notion was dubbed Sturgeon's law. Given that Sturgeon played guitar and specialized in "bawdy" songs, it's unlikely that he would have rated a disco song like Gwen McCrae's "90 Percent of Me Is You" among that rare 10 percent. Sturgeon's law might be cranky, but it has a crucial upside: The worthy 10 percent—not just in science fiction, but in every artistic endeavor—makes human expression the vibrant, life-giving force it is. Truly, even Sturgeon would have agreed that if we can simply free our minds, our asses are sure to follow.

THE VERDICT

Guided By Voices became certifiable indie sensations with *Bee Thousand* in 1994. The following year, signed to a lucrative deal with Matador, Robert Pollard and his mates were no longer shackled to the lo-fi, four-track setup they'd worked with for nearly a decade, by necessity as well as aesthetics. But with *Alien Lanes* already in the can, the band insisted it be released in all its un-pristine glory—newfound access to a recording budget be damned. "The cost for recording *Alien Lanes*, if you leave out the beer, was about ten dollars," wrote James Greer, who played with the band and chronicled his experiences in *GBV: A Brief History*. In retrospect, the gain from that $10 investment is incalculable.

"Blimps Go 90" is tucked away deeply into *Alien Lanes*, track 22, and from its ordinary but somehow mythic sounding title on down, it's a thing of brief, unfussed-over beauty. "Blimps" follows the ominous, similarly aeronautically minded "Striped White Jets" with a dreamy pastoral vibe via Greg Demos's toddling violin and the comforting simplicity of the main guitar riff. Pollard's vocal melody is casually insinuating, singsong but with a trace of grit. His imagery rolls by like Super 8 footage, combining personal mythos with a tinge of weird old America. Trippy doggerel mixes with Beat poetics ("Weep sad freaks of the nation"). Most telling is what seems to be a reference to his former work as an elementary school teacher: "Oftentimes I'm reminded of the sweet young days/when I poured punch for the franchise."

GWEN MCCRAE: "90 Percent of Me Is You"

PREFUSE 73: "90 Percent of My Mind is With You"

If 6 Was 9

But let's not overdo the search for meaning here. Like many of Pollard's songs, "Blimps Go 90" began with a title, one that would look good skywritten or presented in sequence, like a Burma Shave sign. Pollard's peculiar genius is coming up with music so well suited to the fantasy titles his mind keeps churning out. In an interview around the time of *Alien Lanes*, he said, "Titles are what used to make me buy records as a kid. If I wasn't too familiar with the band, and they had a good band name, and the titles were cool on the record, and it had a fairly neat cover, I'd buy it."

That said, Pollard's title is no mere bit of fluff—it's an actual statement of fact. As Wikipedia notes, "Given the large frontal area and wetted surface of an airship, a practical limit is reached around 80–100 miles per hour."

In other words, blimps go 90.

On with the show…

A SONG TITLE	AN ALBUM	A BAND	A LYRIC
"90 Miles an Hour (Down a Dead End Street)" – Hank Snow (1963)	*90* – 808 State (1989)	Ninety Pound Wuss – Port Angeles, Washington, experimental punk (1990s)	"Ninety miles an hour, girl/ is the speed I drive" – Jimi Hendrix, "Crosstown Traffic" (1968)

91

The Bon Savants fused a sophisticated pop sensibility with an appreciation of noise and chaos in an identifiably '00s-era amalgam. The band's best-known single, the irresistible "Between the Moon and the Ocean," emphasizes catchiness, while the hard-crunching "91" offers a piquant glimpse into the workaday reality of the indie world, the pressure to be of the moment, and how quickly that moment changes. Co-founder Kevin Haley provided details of the song's origins to me in a 2014 email.

"I-91 is an interstate in western Massachusetts. When Thom [Moran] and I started the Bon Savants, I was living in southwestern Connecticut and would drive up I-91 to his house in Northampton, Mass., on most weekends to work on music, etc. A few years later, we had relocated to Boston, working our way up to larger rooms and soon found ourselves part of a scene of like-minded bands. However, being wary of the disingenuous nature/ glad-handing that goes with that, I retreated to my apartment way across town, surfacing only to play shows and rehearse.

"At the same time we were getting confused for being "another Interpol," and dance-punk was showing up as a fashionable trend for bands, which I was deeply suspicious of. I wrote "91" as a commentary on this, using a reference to Gang of Four as a sarcastic jab at how elements of 70s/ 80s post-punk had been repurposed as a sound du jour for groups who months earlier were still writing third-rate Travis songs. All I wanted to do was make music, hence the chorus 'that's it I'm going home,' plus the 91 in the title as a call for heading back to what made being in a band enjoyable in the first place."

THE VERDICT

Somewhere in that nether zone between movements in the nondescript late '80s came Mudhoney: high-intensity sons of the Stooges whose molten stew of punk, psych, and metal was the touchstone element of the Seattle music scene and the sound that, for better or worse, came to define it. "I saw Nirvana as a four-piece in 1989, not long after I'd seen Mudhoney," recalled Steve Shelley of Sonic Youth. "I remember thinking, 'these guys are pretty good, but they're no Mudhoney.'"

Mudhoney arose out of Green River, who were named not for the CCR song but for a then-at-large regional serial killer. The band broke apart when one faction (including future Pearl Jammers Jeff Ament and Stone Gossard) sought mainstream respectability, and the other—singer Mark Arm, guitarist Steve Turner, and drummer Alex Vincent—said fuck that and became Mudhoney.

In a spirit of inclusivity, Mudhoney drew on a slew of worthy influences: the gritty garage-psych of local legends the Seeds, the brute force of Blue Cheer, the fuck-off attitude of punk rock. And surf music. And Russ Meyer films, of course. These untrendy elements reflected the well-honed aesthetic of Mark Arm, whose previous projects included the Thrown Ups, makers of what he called "music that sounds like vomit looks."

More than anything, Mudhoney reveled in fuzz, which has a long and proud history that's worthy recalling. You can find it in isolated spots in the early 1960s, such as the Ventures' "2000 Pound Bee," but 1965 was ground zero for the stuff. "Satisfaction"—made so memorable by Keith Richards' Gibson Maestro Fuzztone, one of the first commercially available fuzz boxes—topped the charts. On *Rubber Soul*'s "Think For Yourself," Paul McCartney is credited with "fuzz bass." The world had not seen that kind of credit before.

But the fuzz craze didn't last. An increasingly clean and dry production aesthetic prevailed in the next decade, and when fuzz returned it was one ingredient among many in a stew of overdrive and distortion. I can't prove it, but somewhere around 1988, when Mudhoney released *Superfuzz Bigmuff*, fuzz came storming back as a glorious end in itself. Three years later they were on a major label. As the sole instrumental on 1991's *Every Good Boy Deserves Fudge*, "Fuzz Gun '91" lacks the gleeful, unhinged yowl of front man/chief visionary Mark Arm and may not qualify as an essential Mudhoney track, but the band's sweaty intensity is impossible to miss and its allegiance to a certain distortion pedal never wavers. On the strength of a knotty power chord riff, Steve Turner's face-melting squiggles, and Dan Peters's emphatic drumming, it's compelling testimony to the enduring appeal of wordless riff rock, played fast and loose and flambéed in fuzz.

A SONG TITLE	AN ALBUM	A BAND	A LYRIC
"91" – Death By Stereo (2001)	*Apocalypse 91…The Enemy Strikes Black* – Public Enemy (1991)	Ninety One – France, electronica (2000s)	"I been a freak since '91" – Azealia Banks, "P-U-S-S-Y" (2012)

92

Ninety-two has one crucial moment in the musical sun, courtesy of "The Christmas Song"—you know, the one with the chestnuts roasting on an open fire. "And so I'm offering this simple phrase to kids from 1 to 92," begins the final chorus sung by Nat Cole and Justin Bieber and a zillion others. It may well be the most performed yuletide ditty of all, if you can trust BMI. Fountains of Wayne's Xmas song "I Want an Alien For Christmas" may not be as memorable, but their "92 Subaru" is a rare bird indeed: an authentically triumphant car song extolling the virtues of a ride that also happens to get excellent gas mileage. Someday I'd like to cruise in that late-model baby blue breezer while cranking "Damn 92" by Las Robertas, an all-woman Costa Rican indie pop band enamored of the Breeders and the C86 bands, who do honor to their heroes.

THE VERDICT

Siouxsie & the Banshees didn't form so much as they were willed into being. A slot opened up in London's Punk Rock Festival in September 1976; Siouxsie Sioux said she had a band, and few days later she did. It included future Adam & the Ants guitarist Marco Pirroni and Sid Vicious. They performed a fully improvised set interpolating the Lord's Prayer and Bay City Rollers covers.

Up to that point, Sioux had been the best-known insider in the Pistols entourage, at the forefront of the "Bromley contingent," named for the stifling London suburb its members had called home. She was as much as a provocateur as Rotten-Lydon. Bare breasts and a swastika armband were her typical ensemble in the primal punk days. After the demise of the Pistols, those clamoring to fill the void found it was harder than it looked. As Jon Savage wrote in *England's Dreaming*, "Punk's accessibility seemed to suggest that anyone could do it: the Roxy [the hub of live punk music in London c. 1976] proved that not everybody could." One would seriously doubt the smart money was on Siouxsie and company. But with the release of 1978's *The Scream* the band toured the U.K., serving notice that a Valkyrie had arrived with the musical imagination to match her siren voice.

In 1986, eight years and eight LPs into their career, Siouxsie & the Banshees had weathered lineup changes, challenged themselves with new stylistic avenues, and were in danger of becoming elder statesmen—a most un-punk thing to be. Not that Sioux was concerned. "It was meant to be our 15 minutes of fame," she said at the time, "but we managed to sustain it for 10 years, which only goes to show how addictive dressing up and making noise for a living can be."

"92 Degrees," from *Tinderbox*, originated with a news article about the phenomenon of increased crime brought on by extreme temperature. Beginning with a clip from the 1953 B-movie *It Came From Outer Space*, in which a detective posits the same theory, the song evokes the early stages of panic as the temperature rises and violence becomes a certainty. The snippet foreshadowed Sioux's growing fascination with sampling, which would send her in a dancier, less guitar-based direction in the next decade.

Here the sound is grand-scale, as if emanating from a huge stone room. Opening with a quasi-military snare pattern, guitarist John "Valentine" Carruthers unspools harp-like figures that fill up the void with a subtlety that John McGeoch (whom Carruthers ostensibly replaced) would have admired. Carruthers adheres to Sioux's long-held demand that a guitar not sound like a guitar and fulfills the textural role that distinguished the posthumously lauded McGeoch. The production has more polish compared with previous records, and Siouxsie, as is her wont, unsettles us, her warning both captivating and merciless.

A SONG TITLE	AN ALBUM	A BAND	A LYRIC
"92" – Tracy Walton (2011)	*92* – Charles Aznavour (1992)	Summer of '92 – Toronto, Celtic-punk-folk rock (2000s–2010s)	"And he stuck it in his collection/ Section 92" – Frankie Lymon, "Goody Goody" (1957)

93

Lorena Bobbit's memorable act of revenge and the suicide of Clinton White House counsel Vince Foster dominated the headlines during the American summer of 1993. Among several songs named for this period, I'm most keen on "The Summer of '93" by Bobby Bare Jr., an under-the-radar singer-songwriter and the son of the country crooner who had hits with novelty songs like "Dropkick Me Jesus (Through the Goalposts of Life)" as well as traditional folk and country fare. Bare Jr. doesn't dismiss country, but he adds the "alt" prefix defiantly. Backed by three-quarters of My Morning Jacket, he nabs the M.O. of that band's "Wordless Chorus" while using the verses to paint a very specific memory: "I saw them with Sonic Youth in the Summer of '93//The singer spit while he screamed//With broken cowboy boots wrapped around his crooked feet." After some snooping around, I found that the band in question turns out to be Jesus Lizard, who were notorious for chaotic live shows.

THE VERDICT

A decade after Mark "Stew" Stewart issued the first of six critically lauded records, his semi-autobiographical theater piece *Passing Strange* made it to Broadway. Based on a lifetime of struggling to fit into various ill-fitting subcultures, it earned him long-overdue name recognition, despite two of his records having already received top honors in the year-end poll of the exceedingly mainstream *Entertainment Weekly*.

Yet it's not really surprising that Stew toiled in the shadows for so long. Working in a hard-to-pin down amalgam of styles he dubbed variously as "Blacharach" and "Afro-baroque," wary of careerism, operating under the toxic tongue-in-cheek band name the Negro Problem, Stew's made no secret of his distaste for the star-making machinery. That sentiment was clear on his debut album, *Post Minstrel Syndrome*, which began by questioning the bona fides of *L.A. Times* music critic Robert Hilburn before moving onto the deceptively sunny "If You Would Have Travelled on the 93 North."

While "93 North" exemplifies Stew's "Blacharach" vein—a nod to another writer fond of limning the lambent surfaces of L.A.—it also expresses a darker view of the city that recalls the songs of Love's Arthur Lee, whom Stew supported on his 2003–04 comeback tour. Lee had a similar way of evoking both the light and shadow. One of his songs begins, "Sitting on a hill-*side/ watching all the people…DIE*," and there's a similar jarring dichotomy as Stew restlessly veers between the beauty of the road not taken and the soulless technology that purports to keep us connected but only preserves our personal bubbles.

"93 North" embodies both the glory and contradictions of Stew in one tidy package. On the one hand, it's scary-catchy, ebullient even, but from a title in the imperfect subjunctive voice, it's fraught with ambiguity, willfully complex and open-ended. Starting with fanfare suggestive of mariachi-as-done-by Frank Zappa, it rolls along for two bars at an easy-going, Byrds-like tempo. Then, thrillingly, Stew leaps up to an urgent almost-falsetto that catapults us into the sky above that sun-dappled highway. Just as quickly, he yanks us back to the reality of prefab L.A.: "Cellular phone/ in a cardboard home,"[1] he croons, each line echoed by his partner/bandmate Heidi Rodewald, the show-music uplift at odds with the sour sentiment. At the two-minute mark there's an exquisite jazz-chord-laden segment followed by the two main sections presented in gorgeous counterpoint. But rather than leave us purely exhilarated, Stew brings back that skewed mariachi riff and ends things on a determinedly unresolved note.

A SONG TITLE	AN ALBUM	A BAND	A LYRIC
"93 Avenue Blues" – Swans (2012)	*Xmas 93* – Saint Etienne (1993)	Current 93 – England, experimental (1970s–2000s)	"She sang it as she tucked me in/ When I was 93" – The Goons, "The Ying Tong Song" (1957)

---

[1] Cellular phones first began appearing in song lyrics in the late 1980s—one early mention was in Sir Mix-A lot's "My Posse's On Broadway," but "93 North" seems to be the first to cast the technology not as a desired material good but as symptomatic of our late-20th century anomie.

94

Straddling the rock/avant-garde/experimental worlds, Jim O'Rourke has been a member of Sonic Youth, a producer for Wilco, a film scorer for Werner Herzog, a prolific and sought-after collaborator, and a maker of accomplished, diverse solo recordings. In the late '90s O'Rourke made a string of three records[1] in the style and spirit of John Fahey, one of his heroes and eventually a personal friend, who similarly explored traditional American musical forms while embracing experimentalism and electronic instrumentation. "94 the Long Way" is one of four lengthy compositions on *Bad Timing*, a 1997 release that reflected O'Rourke's longtime interest in improvisation. Beginning with a halting acoustic guitar cadence, it blooms midway with heavenly slide guitar, brass, and piano accompaniment into a transporting pastoral reverie.

THE VERDICT

At the forefront of the mid-1970s Australian underground were the Saints, of Brisbane, and the Sydney-based Radio Birdman. The Saints made a name for themselves with the incendiary "I'm Stranded" and soon relocated to London. Radio Birdman, named for a misheard Stooges lyric, stayed in Sydney, winning the hearts and minds of the next generation of punk rockers and becoming unofficial kings of the indie demimonde. As Mudhoney did in Seattle in 1988–89, Radio Birdman became the linchpin of a scene that inspired people to believe that they, too, could pick up instruments, form bands, and make music of their own.

Radio Birdman started with Deniz Tek, a skilled guitarist and songwriter from Detroit who abandoned his medical studies in Sydney to pursue music. After serving time in an acrimonious, rough-and-tumble outfit called TV Jones, he hooked up with Rob Younger, a surfer and a hellraiser who had no trouble slipping into the role of gyrating, shirtless frontman. "Deniz and I were very close," he recalled in an interview, "and we were pariahs. We went out, and wherever we turned up with that band, we caused some shit. To me it seemed like that had meaning in itself."

With the addition of a few more band members, including keyboardist Pip Hoyle (another medical student) for a Doorsian flavor, the lineup was complete. Together they made primitively throbbing, unhinged rock in the aggressive tradition of Tek's native Detroit region: the Stooges and MC5 crosscut with the jammier instincts of another northeastern band, Long Island's Blue Öyster Cult. Tek was the prime force, the main songwriter and aesthete. Just as Mark Arm brought his love of the Seeds and '60s trash, Tek brought with him the musical gods and landscape of the Motor City.

---

[1] Each was named after a film by Nicolas Roeg.

If 6 Was 9

"I-94" had been kicking around since Tek's days with TV Jones. Originally titled "Eskimo Pies," it was the planned A-side of a single recorded in March 1974 that never saw the light of day. Recorded in 1978, shortly before the band broke up, it was released three years later on *Living Eyes*. Tek told author Vivien Johnson that he wrote the song after experiencing extreme withdrawal symptoms for his beloved Detroit—and its ice-cream treats. To be sure, it's chock-full of Michigan-iana, but it's a strange kind of miss-you song. "Eskimo Pies comin' to you" sets you up for something quite apart from "Burning to you straight from hell." And the bit about, "You can go to Europe with Jean-Paul" is a left-field Francophile head-scratcher lyric to rival Scott Walker's "Shaking hands with Charles DeGaulle," with which it rhymes.

A SONG TITLE	AN ALBUM	A BAND	A LYRIC
"United '94" – Psychic TV (1994)	*94 Diskont* – Oval (1995)	94 East – Minneapolis, funk (1970s)	"…round and round the hotel garage/Must have been touching close to 94"– David Bowie, "Always Crashing in the Same Car" (1977)

95

As purveyors of short, tuneful blasts, the Ramones had few equals. Most Ramones classics clock in under two minutes. But for some reason, when asked to identify the absolute shortest Ramones song, the pundits-in-training who supply Answers.com with answers fumbled the ball, claiming the honor belonged to "Now I Wanna Sniff Some Glue" at 1:28. They were wrong by a large margin. The shortest Ramones song, and as far as I can tell, the only song attributed solely to Johnny Ramone, is "Durango 95," an instrumental named for the make of Alex De Large's beloved car in *A Clockwork Orange*, which "purred away real horrorshow – a nice, warm, vibraty feeling all through your guttiwuts."

THE VERDICT

Year-of-release provides the framework for *'64–'95*, a concept album by the English duo Lemon Jelly on which each of the nine proper tracks is principally based on a single sample. A purist might approach this framework by selecting music that typifies the sound of the year in question, but Lemon Jelly—that's Fred Deakin and Nick Franglen—have from the beginning avoided the orthodoxy that afflicts many crate-digging DJ types. Instead, they've shown a predilection for left-field vocal samples, like sputtering outer-space-talking astronauts and a chortling tenor rhapsodizing about ducks. Key sources for *'64–'95* range from a drone-rocker by the short-lived Edinburgh band Scars to "square" '60s fare like Cambridge, England's King's Singers and New Zealand pop crooner John Rowles.

First off: a hat tip to Messrs. Franglen and Deakin. With *64–'95*, Lemon Jelly joins the elite company of Eno's *Music For Airports*, Karma to Burn's *Wild Wonderful Purgatory*, and the Postmarks' *By the Numbers* in making a full-length LP consisting of nothing but number songs.[1] Further, the songs amount to a godsend for the intrepid numerologist geek who frets away his nights cogitating over those hard-to-fill later digits. Lemon Jelly positively *glory* in the crooked numbers: There's "'79 aka The Shouty Track." There's "'93 aka Don't Stop Now"—we're talking surpassingly crooked here. But let's be reasonable. We'll only go to the Well of Lemon Jelly once, and now is the time to do it. Deploying the Jelly at 95 does two crucial things: it fills a particularly barren quantity exceedingly well, while spotlighting the record's most blissful moment.

The first two Lemon Jelly releases were as softly alluring as the neon-hued cartoon worlds of Deakin's LP covers, but the mood darkened considerably on *'64–'95*, as on the aforementioned "Shouty Track" and the paranoid William Shatner narration on the

---

1   Technically, the album kicks off with the 24-second "It Was …" in which a heavily accented man, over an orchestral fragment, attempts to summon a date ("That was in 19…uh…19…let me see…") But we won't count that.

thunderous closer. Smack in the middle of it all lands "'95 aka Make Things Right," and when it does, the clouds part and the sunshine is bright and glinting with tracers.

The source song here is "Before You Walk Out of My Life" by Monica, the 14-year-old R&B phenom who hit No. 1 with her first two singles and still holds the distinction of being the youngest person to do so. Franglen and Deakin grab two crucial bars from the refrain ("Only wanna make things right") and, adding their trademark acoustic arpeggios, string it into an endless chorus.[2] The choice not to use Monica's preternaturally assured vocal but rather to enlist English R&B singer Terri Walker to re-sing the line and scat over her own looped vocal might have been economical or aesthetic in origin. In any case, "'95 aka Make Things Right" is pure bliss, from its dreamy fade-in through the entrancing, unexpected bass-led sequence right before the sun slips below the horizon.

A SONG TITLE	AN ALBUM	A BAND	A LYRIC
"I-95" – Fountains of Wayne (2007)	*Incidental Music 1991–95* – Superchunk (1995)	95 South – U.S., rap (2000s)	"Who's coming home on the old ninety five?" – Buffalo Springfield, "Nowadays Clancy Can't Even Sing" (1966)

---

[2] Moon Guitars of Glasgow makes Lemon Jelly's preferred acoustic guitar. The moon guitar is a variety of Chinese lute, properly called a yueqin, with an orb-shaped soundboard.

96

In "1865 (96 Degrees in the Shade)," Third World reanimates an ugly episode in Jamaican history and weds it to a deeply singable song that does not demand that the listener be brought down by the events in question. The subject is Jamaica's Morant Bay Rebellion of 1865 and the martyrdom of George William Gordon. After putting down a small uprising, colonial governor Edward John Eyre ordered the arrest of Gordon, a political enemy who had nothing to do with the violence. After a sham trial, Eyre had Gordon hanged publicly, with great pomp and circumstance. Gordon is now hailed as a national hero of Jamaica.

THE VERDICT

"96 Tears" by ? and the Mysterians topped the American pop charts almost exactly a century after the events of Morant Bay, and it too is associated with rebellion, albeit of a less cataclysmic sort. It is very likely the most influential entry on this list. Quite a claim, I know, but think about it. Sure, "Eight Days a Week" offered object lessons in the art of the three-minute pop song that would hit home with countless songwriters, including the young Kurt Cobain, but so did many Beatles songs of that period. "19th Nervous Breakdown" surely embodies the Stones' unmatched mid-'60s stride, but so does "Satisfaction" and "Get Off of My Cloud." On the other hand, "96 Tears" is in a category all its own.

Attesting to its singularity are several factors. In a broad sense, the song is the standard-bearer for first-wave American garage rock, a genre that remains vital today. Where would the Black Keys be without those underlying garage-rock roots? "96 Tears" is unique in taking a number devoid of intrinsic meaning or associations and imbuing it with a lasting identity. Try this simple word-association test: Say "96" to another person and see if the first response isn't tears. Does any another numeral do that? "96 Tears" is one of the first No. 1 songs by a Latino rock band. Then consider the staggering number of cover versions the song has spawned, ranging from Suicide to Thelma Houston. Of the songs I've considered thus far, only "Route 66" rivals it for the sheer diversity of cover versions. The song's central image has become so embedded in the vernacular that it has been referenced numerous times in different eras, like the "96 tears through 24 hours" nastiness at the core of X's "Johnny Hit and Run Paulene," the "97th tear" in Okkervil River's numerically obsessed "Plus Ones," and "Human Fly" by the Cramps: "I'm a human fly and I don't know why/ I've got 96 tears in my 96 eyes." On the subject of eyes, "96 Tears" was put to fiendish good use in *Stephen King's Cat's Eye*.

Embodying the song's menace is the former Rudy Martinez, the band's lead singer, who legally changed his name to the question mark symbol and has remained outlandish and eminently quotable since the world first met him in 1966. Appearing in public always in sunglasses, he stoked the mystery of his identity during his heyday. In interviews he claimed to have no real musical influences and had very little good to say about his competition. Yet despite his eccentricity and contrariness, the man has been surprisingly gregarious when discussing his best-known creation. The song's dominant organ sound, he has said, derives from his Catholic upbringing, the propulsive beat from the enraptured singing and handclapping he'd hear coming home from Sunday Mass as he passed the local Baptist church.[1]

And as for that ineffable, unremitting sense of dread?

"Remember the scene [in *Gone With the Wind*] with the little girl falling off the horse, that terror theme? I wanted that *terror* in there as well."

The singer's explanation for choosing 96 was simpler: "It had a magic ring to it."

A SONG TITLE	AN ALBUM	A BAND	A LYRIC
"96th Street School" – Red Norvo (1957)	*If 69 Was 96: Pinguin Moschner and Joe Sachse Play the Music of Jimi Hendrix* (1994)	U96 – Hamburg, Germany, techno (1990s)	"Don't you blame us/ You 96-decibel freaks" – Mott the Hoople, "The Golden Age of Rock 'n' Roll" (1974)

---

1 Further to 96 and church music, the Hammond L 100 electric organ, first sold in 1961, achieved a sound akin to that of a church organ via 96 metal tone wheels interacting with 96 electromagnetic pickups.

97

Scalded to Death by the Steam, a book-length exploration of the train-wreck ballad by Katie Letcher Lyle, takes its title from a line in the country standard "The Wreck of the Old 97." Lyle describes the strict formula of these songs, which begin with a bright, hopeful future in the first verse, a religious homily in the last, and tragic human error in between. She also reminds us that not all of these were based on actual events. "The Wreck of the Number Nine," for example, in which the dying engineer gets a message to his beloved, saying he's built a house for them to live in, was the made-up kind. But "The Wreck of the Old 97" is the real thing.

It commemorates a hellish crash in September 1903, when the time-pressured engineer of Southern Railway locomotive 1102 was moving too fast on a three-mile downgrade to negotiate a sharp turn near Danville, Virginia. The locomotive, along with mail car No. 97, plunged from a 45-foot-high trestle into a ravine and caught fire. The event might have been lost to time had not a ballad been composed soon thereafter by the people who arrived on the scene and tended to the dead. The verses were eventually wedded to a tune derived from "The Ship That Never Returned," a folk ballad written 50 years earlier by Henry Clay Work, which also addressed the perils of the industrial age. First recorded in 1922 by Virginia musician Henry Witter, "The Wreck of the Old 97" was popularized a year later by Vernon Dalhart and became the first million-selling country record.

THE VERDICT

Sometimes you come back and try the same food, and lo, it's better than you remember it. I saw Sleater-Kinney open for Guided By Voices at New York's SummerStage in Central Park in 1996. I was there for Pollard and the boys, who obliged by kicking out the jams and chucking full cans of Miller into the far reaches of the audience. (Robert Pollard, after all, was a decorated high school athlete who threw a no-hitter as a member of the Wright State University baseball team.) Truth be told, though, I was not feeling especially open-minded about the opening act, a trio of women from Seattle who were making a hell of a racket up there. Singer Corin Tucker's voice is inevitably described as "love it or hate it." I loved the tight playing and the self-possession, but I didn't become a fan, despite the heavy breathing the band has engendered among rock critics and devotees throughout its existence.

When my friend and colleague Jeff Klingman—a Portlander and a longtime Sleater-Kinney lover—said he had a #97 song for me, I wasn't enthusiastic, given my initial resistance to the band. But I put it on and was floored immediately. What had kept me from them? Were all their songs this good? Jeff offered an explanation.

"I think "Dance Song '97" is a gateway drug into that band, the mix-tape track," he wrote in a 2014 email. "They can be kind of intimidating, a lot of anger even when they are being funny, like 'I Wanna Be Your Joey Ramone.' So, the fact that the beat and those great little minor-Television guitar riffs back and forth are kind of confident and laid back, it lets you into Corin Tucker's vocal in a way that a lot of their other stuff didn't. It was the first sort of moment when you felt like they didn't have a chip on their shoulder, so for someone who might not have a lot of reason to identify with them personally (a HUGE part of their appeal, I'd argue) you could get into it without feeling like it wasn't made FOR you getting in the way."

Of course, these nuances mean nothing to the hardcore fan. Lianne Saussy, 12, and her sister, Caroline, 10, my neighbors in Chapel Hill, required no gateway. They have been hooked on Sleater-Kinney since their dad, David, changed up his automotive playlist a few years ago. He had been struck by the fact that most of his preferred music was made by men, and decided it would be good to expose his daughters to this defiant, self-possessed all-woman band. The girls just about went nuts in their car seats the first time he popped *All Hands on the Bad One* into the cassette player. It remains their favorite LP—because of songs like "You're No Rock N' Roll Fun" and "Ballad of a Ladyman"—but they do like "Dance Song '97" a bunch.

"I like the way she sings to the lyrics," says Lianne.

Caroline nods and grins.

A SONG TITLE	AN ALBUM	A BAND	A LYRIC
"97 Lovers" – Pulp (1996)	*Une Rentrée 97* – V.A. (1997)	Old 97s – Dallas, alt-country ('90s–2010s)	"There are ninety-seven sides to everything" – Close Lobsters, "Pathetique" (1987)

98

J. Snowden is an African-American schoolteacher from Billerica, Massachusetts, who found her creative spark after visiting Canada in 1989. She has been tagged an outsider artist since the term became popularized in Irwin Chusid's *Songs in the Key of Z*, which shone a spotlight on people working on the fringes of the world of artistic expression. It featured people like Jack Mudurian, a nursing home resident who earned a cult audience with his continuous rendition of 129 American songs, along with better-known oddball phenomena like the Shaggs and Daniel Johnston. Though Ms. Snowden emphatically rejects the "outsider artist" label, she fulfills its essential prerequisite in her complete unselfconsciousness and lack of ulterior artistic motive.

She holds a degree from Berklee College of Music, but there's nothing polished about her songs or presentation—her voice is pitchy and her melodies are not deep. Yet this is not kitsch, irony, or weirdness for its own sake. She writes and plays songs simply to bring joy—and she succeeds.

Ms. Snowden may not have been thinking of "Rocket 88" when she wrote "98," but both songs celebrate the power of Oldsmobiles and the allure of speed. In the more famous song, Jackie Brenston praises his car's "modern design," while Ms. Snowden talks about "good standard equipment." Unlike "Rocket 88," and car songs in general, "98" isn't freighted with attendant themes of escape, intoxication, or backseat activities. There is no subtext at all: "I can hardly wait/to buy a 98//Put your foot on the gas/ and you can go fast." And yet the song culminates with a final, genuinely experimental touch—an ending chord taken from orchestral performance, followed by applause—reminding us that beneath the jaunty car toots and engine noises, a curious, artistic sensibility abides.

THE VERDICT

As the New Year dawned in 1967, the charts were so bright with sunshine pop you almost had to wear shades. The Monkees' "I'm a Believer" floated to the top chart position and lingered into February, setting the stage for a Philadelphia singer named James Barry Keefer to have his moment of glory as the mono-monikered Keith.[1] In "98.6," which peaked at #7 on the *Billboard* Hot 100, one can detect a definite Cheshire cat grin that was right in tune with the cannabinoidal times. John Lennon liked it so much that he complimented Keith when they found themselves side by side in an L.A. nightclub's men's room. Oh, to have been a fly on that

---

[1] Sunshine pop seems to have engendered a trend toward bland, single-named pseudonyms. Besides Keith there was Derek, a Scottish recording artist born John Hendry Blair who had a semi-hit with "Cinnamon" in 1968. He also recorded as Brother John, The Eye-Full Tower, and The Non-Conformists. Blair's biggest hit came early on with "Mr. Bass Man," under his best pseudonym: Johnny Cymbal.

wall. And while "98.6" is undeniably lightweight, it's a solid piece of work, right down to the (uncredited) background vocals by the Tokens, who had scored big with "The Lion Sleeps Tonight" a few years earlier.

Keith followed up "98.6" with an exquisite cover of the Hollies' "Tell Me to My Face," and his songwriting successes include "Lazy Day" for fellow sunshine poppers Spanky & Our Gang. He was no mere one-hit wonder. But just as sure as the human body temperature tends to hover around 98.6, Keith will be remembered solely by one song. If his website, 986keith.com, is any indication, the briefly celebrated Mr. K has made peace with that reality.

What's pretty fascinating is that three of the backing players on "98.6" have contributions that extend far beyond this lovely but admittedly ephemeral bit of ear candy. The achievements of Bobby Gregg—who began in the 1940s with bandleader Paul Whiteman, had a Top 20 hit in 1962 with "The Jam, Pt. 1," and drummed on "Like a Rolling Stone" and "The Sound of Silence"—are far too manifold to recount further. Bassist Joe Macco, a valued session man (that's him on Tommy James' "I Think We're Alone Now") was a member of the famed MFSB crew, which anchored the Sound of Philadelphia during the early 1970s. Guitarist Vinnie Bell, who like Gregg was a session player at Philly's influential Cameo-Parkway Records, invented one of the first electric 12-string guitars and certainly the world's first electric sitar. Some of his ideas didn't take, like the bouzouki he once electrified, but Bell came up with several notable innovations. His "pedalboard tremolo," for example, is responsible for the memorably rubbery bass sound in Angelo Badalamenti's "Theme From Twin Peaks."[2]

A SONG TITLE	AN ALBUM	A BAND	A LYRIC
"Bondi 98" – Dick Diver (2013)	*S.D.Q. '98* – Doug Sahm (1998)	98 Degrees – L.A., R&B-pop, (1909s–2010s)	"An old man turned ninety-eight/ He won the lottery and died the next day" – Alanis Morissette, "Ironic" (1995)

POSTSCRIPT: James Barry Keefer, formerly Keith, now goes by Bazza Keefer, in tribute to his mother.

---

[2] Vinnie Bell can also lay claim to inventing "the watery guitar sound," which is to Muzak what the power chord is to rock 'n' roll. Eventually the watery guitar sound would become cool again, thanks to the Cure, but for a while it was strictly Squaresville.

99

The plan was always to stop at 99. As I imagined it, the journey would culminate in one final, joyous release of *luftbalons* and a triumphant, complete rendition of "99 Bottles of Beer on the Wall." But when Nena's massive hit from 1983 came up on Shuffle recently, an odd thing happened. Usually I'll let anything play when I'm in the middle of cooking dinner, but this time, acting purely on instinct, I wiped off my food-covered hands just so I could press Forward. In that moment I knew the #99 slot was up for grabs.

What about Jay-Z, you ask? Well, as Pee-Wee Herman might have put it, "I Like Jay-Z. *Like*." And I do like "99 Problems" more than most within the Hova canon. But the final song on this epic journey has to be one I can extol, champion, proselytize on behalf of, bellow from the highest building. Discovering that there is such a song has given me solace and many nights of peaceful slumber.

This climb began with a myriad of choices, so it feels right that our endpoint is also a bit of a monster. For one thing, 99 has a luscious assonance that conjures the clangor of tolling bells. Listen to Ray Davies linger in "Autumn Almanac" ("If I live to be *naa-ah-ahn-ty nine...*") or check out verse 1 of Dwight Twilley's signature hit, "I'm On Fire," for a slapback-echo-enriched, falsetto take on the number. No doubt, 99's sheer sing-ability is part of its appeal.

In American song, 99 is firmly associated with a life sentence. In prison songs like "Send Me to the 'Lectric Chair" and "Junker's Blues,"[1] judges punish those they find guilty with 99 years in jail. It also pops up in murder ballads ("Delia") and traditional bluegrass tunes, such as "Ninety Nine Years and One Dark Day," "99 Years Is Almost For Life," and the traditional "I've Still Got 99," the title of which Woody Guthrie incorporated into his "Worried Man Blues." Still, the phrase served as a declaration of undying love and loyalty in "Black Angel Blues" by Lucille Bogan: "Womens don't bother my black angel: don't bother him in any way//"I'll serve ninety-nine years in jail: most any day."

The murder weapon in these 99-year songs is inevitably a .44, and the songs often end up in court. In one of the most famous, "Ninety Nine Year Blues" by Julius Daniels, the singer implores, "Be light on me, judge, ain't been here before," and the judge thunders back, "Give you ninety-nine years, don't come back here no more." In "Ninety-Nine Year Blues," which Jimmie Rodgers popularized in 1932, the judge is just as cruel: "Boy, you got two sixes/they're all upside down." In Bruce Springsteen's "Johnny 99," which draws considerably on the Julius Daniels song, the accused asks the judge to put him out of his misery rather than put him away for life.

Ninety-nine is also tightly bound to 99 percent, that maddeningly just-shy-of-total quantity often wedded to a complaint about the missing shred. "Ninety-Nine and a Half,"[2] written and popularized by the self-described "raggedy voiced" gospel singer

---

[1] According to Dr. John, "Junker's Blues" (variously adapted by Lloyd Price, Professor Longhair, Fats Domino and the Clash among others) was "the anthem of the dopers, the whores, the pimps, the cons" and its rhythm was known as "the jailbird beat."
[2] In *Sacred Music*, his controversial conflation of Christian texts with jazz from 1966, Duke Ellington borrows the lyrical essence of the Coates song for a composition called "99%."

Dorothy Love Coates in the 1950s, pivots on the notion that when your goal is 100, "99 and a half won't do." Little Richard was an enthusiastic fan of Coates and incorporated some of her vocal mannerisms into his own style. He covered "Ninety-Nine and a Half," as did the Supremes, who had a massive hit with a song that borrowed and reworked the title of another Coates nugget, "You Can't Hurry God." Cropper, Floyd & Pickett lifted Coates's phrase for the R&B standard "Ninety Nine and a Half (Won't Do)," which has proved versatile enough to please Creedence, Springsteen, and the Trammps.

THE VERDICT

Along with the 13th Floor Elevators, the Moving Sidewalks were the edgy people-movers of the Texas rock scene of 1967. The Sidewalks were closer to the Hendrix school of celestial expansion than were the more eccentric, primitive Elevators. Yet the Sidewalks' biggest hit and greatest song is a stoned garage classic, 2:15 of nerve-jangling rock powered by the keening Hammond B3 organ of Tom Moore and a salacious vocal by Billy Gibbons. Favoring brute force over psychedelic noodling, even as it celebrates the power of drug-induced mind expansion, "99th Floor" starts by copping the oldest blues trope of them all: "Woke up this morning…" But no lament follows. This time, the ancient setup gets a dose of something distinctly 1967: The woman in "99th Floor" isn't treating her man mean. In fact, she's the anti-Eve, selflessly offering him the gift of personal liberation.

Like the 13th Floor Elevators, for whom they opened a few gigs, the Moving Sidewalks met their demise at the hands of that hallowed institution known as Texas justice. At the height of the band's success, keyboardist Moore and drummer Don Summers were arrested on draft-evasion charges (a fate suffered, oddly enough, by the previous chapter's champion, Keith, two years later). Gibbons moved on to create the first edition of ZZ Top, one of the Lone Star state's most successful and enduring bands. His Moving Sidewalks bandmates may not have achieved wider acclaim, but they did live to re-form and tour in 2013 and to appreciate the mark they made with this scorching track.

A SONG TITLE	AN ALBUM	A BAND	A LYRIC
"99" – The Minutemen (1983)	*99%* –Meat Beat Manifesto (1990)	Page 99 – Sterling, Virginia, hardcore punk (1990s–2000s)	"Now I gave my baby now the 99 degree/ She jumped up and threw a pistol down on me" – Robert Johnson, "Stop Breakin' Down" (1937)

ENDNOTE:

The B-side of "99th Floor" is "What Are You Going to Do?" Exactly what I'm thinking.

David Klein

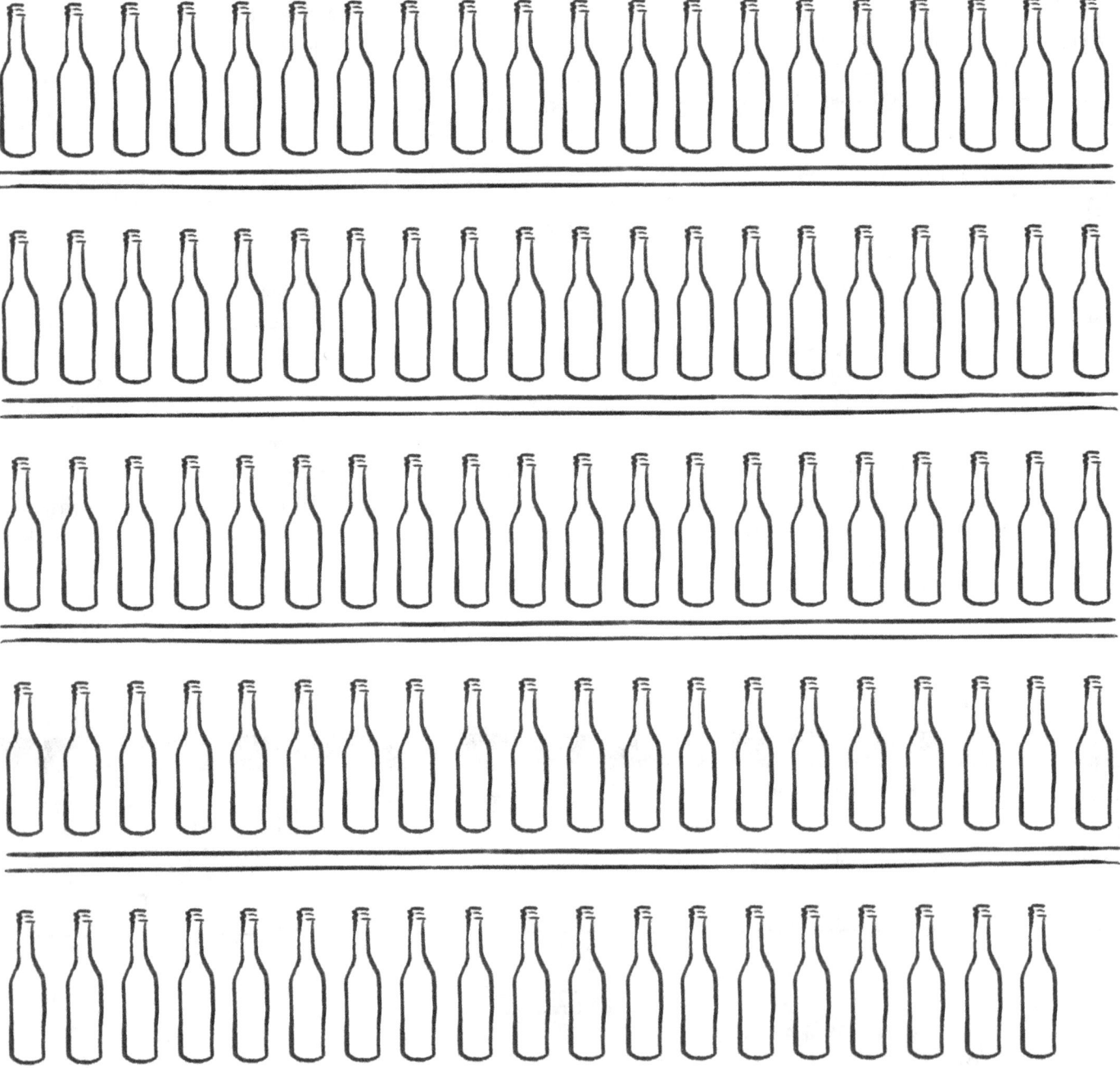

If 6 Was 9

"A Hundred Pounds of Clay" – Gene McDaniels
"107 Steps" – Bjork
"Across 110th Street" – Bobby Womack
"We Are 138" – The Misfits
"ME 262" – Blue Öyster Cult
"365 Is My Number" – King Sunny Ade
"405 Lines" – the Monochrome Set
"Me and My 424" – John Vanderslice
"I'm Gonna Be (500 Miles)" – The Proclaimers
"678" – King Creosote
"712" – Shonen Knife
"720 Times Happier Than the Unjust Man" – The Loud Family
"$1,000 Wedding" – Gram Parsons
"2000 Man" – The Rolling Stones
2112 – Rush
"In the Year 2525" – Zager & Evans
"2541" – Grant Hart
"Three-Five-Zero-Zero" – *Hair* soundtrack
"4,738 Regrets" – Trans Am
5150 – Van Halen
"6,000 Miles From Nowhere" – Snooks Eaglin
"10,000 Words in a Cardboard Box" – The Aquarian Age
"40,000 Headmen" – Traffic
"50,000 Miles Beneath My Brain" – Ten Years After
"The $64,000 Question" – Bobby Tuggle
"100,000 Fireflies" – Magnetic Fields
"1,000,000" – R.E.M.
"Millions" – XTC
"Heaven Is 10 Zillion Light Years Away" – Stevie Wonder

Select Bibliography

Azerrad, Michael. *Our Band Could Be Your Life*. Back Bay Books, 2002.
Breithaupt, Don and Breithaupt, Jeff. *Precious and Few: Pop Music in the Early '70s,* St. Martin's Griffin, 1996.
Browne, David. *Fire and Rain: The Beatles, Simon and Garfunkel, James Taylor, CSNY, and the Lost Story of 1971.* Da Capo Press, 2012.
Buckley, Peter, ed. *The Rough Guide to Rock*. Rough Guides, 1999.
Christgau, Robert. *Christgau's Record Guide: Rock Albums of the '70s,* Ticknor & Fields, 1981.
Cash, Rosanne. *Composed: A Memoir*. Viking, 2010.
Cooper, Kim & Smay, David, eds. *Lost in the Grooves: Scram's Capricious Guide to the Music You Missed*. Routledge, 2005.
Cresswell, Toby. *1001 Songs: The Great Songs of All Time*. Thunder's Mouth Press, 2006.
Dolgins, Adam. *Rock Names from Abba to ZZ Top: How Rock Groups Got Their Names*. Citadel Press, 1993.
Drummond, Bill. *45*. Little, Brown, 2000.
Dunton-Downer, Leslie. *The English Is Coming*. Touchstone, 2011.
Farmer, Neville & XTC. *XTC: Song Stories*. Hyperion, 1998.
Emerick, Geoff. *Here, There and Everywhere*. Gotham, 2007.
Fletcher, Tony. *A Light That Never Goes Out: The Enduring Saga of the Smiths*. Crown Archetype, 2012.
Guterman, Jimmy & O'Donnell, Owen. *The Worst Rock-and-Roll Records of All Time*. Citadel Press, 1991.
Hardy, Phil & Laing, Dave. *Faber Companion to 20th-Century Popular Music*. Faber and Faber, 1990.
Hermes, Will. *Love Goes to Building On Fire: Five Years in New York That Changed Music Forever*. Faber & Faber, 2012.
Freeman, Phil. Ed. *Marooned: The Next Generation of Desert Island Discs*. Da Capo Press, 2007.
Hoskyns, Barney. *Arthur Lee: Alone Again Or*. Canongate, 2002.
Var. A.V. Club writers, *Inventory*. Scribner, 2009.
Johns, Glynn. *Sound Man*. Blue Rider Press, 2014.
Johnson, Vivien. *Radio Birdman*. Sheldon Booth, 1992.
Lazell, Barry. *Indie Hits 1980–1989*. Cherry Red Books, 1997.
Lesh, Phil. *Searching For the Sound: My Life With the Grateful Dead*. Little, Brown, 2005.
Lewisohn, Mark. The Compete Beatles Recording Sessions: The Official Abbey Road Studio Session Notes 1962–1970. Harmony Books, 1988.
Lydon, John. *Rotten: No Irish, No Blacks, No Dogs*. Picador, 1995.
MacDonald, Ian. *Revolution in the Head*. Chicago Review Press, 2005.
McNeil, Legs and McCain, Gillian. *Please Kill Me: The Uncensored Oral History of Punk*. Penguin, 1996.

Marsh, Dave. *The Heart of Rock and Soul*. Da Capo Press, 1999.
Meltzer, Richard. *The Aesthetics of Rock*. Da Capo Press, 1987.
Miller, Scott. *Music: What Happened?* 125 Books, 2010.
Mulholland, Garry. *This Is Uncool: The 500 Greatest Singles Since Punk and Disco*. Cassell Illustrated, 2004.
Paytress, Mark. *I Was There: Gigs That Changed the World*. Cassell Illustrated, 2005
Peellaaert, Guy and Cohn, Nik. *Rock Dreams*. Popular Library, 1973.
Reynolds, Tom. *I Hate Myself and I Want to Die: The 52 Most Depressing Songs You've Ever Heard*. Sanctuary, 2005.
Riley, Tim. *Tell Me Why: A Beatles Commentary*. Da Capo Press, 1998.
Richards, Keith with Fox, James. *Life*. Little, Brown, 2010.
Robbins, Ira, ed. *Trouser Press Guide*. Collier Books, 1993.
The Rolling Stone Interviews: 1967–1980. St. Martin's, 1981.
Scully, Rock with David Dalton. *Living With the Dead: Twenty Years on the Bus with Garcia and the Grateful Dead*. Little, Brown, 1995.
Sheffield, Rob. *Love Is a Mixtape*. Crown, 2007.
Shteamer, Hank. *Ween's Chocolate and Cheese*. Bloomsbury, 2011.
Stanley, Bob. *Yeah Yeah Yeah: The History of Pop Music From Bill Haley to Beyoncé*
Tosches, Nick. *Unsung Heroes of Rock 'n' Roll*. Da Capo Press, 1999.
Wareham, Dean. *Black Postcards: A Rock & Roll Romance*. Penguin Press, 2008.

PERIODICALS
Bananafish
INDY Week
The LA Times
The New York Times
Magnet
Rolling Stone
Texas Monthly

WEBSITES
The A.V. Club
All Music Guide
Billboard
IMDB
NPR
Pitchfork
Internet Archive
Perfect Sound Forever
Songfacts.com
Tisue.net
Wikipedia
YouTube

Acknowledgements

JP Trostle enlivened these chapters with much needed visual pizzazz and was a reliable sounding board for ideas that went far beyond graphics during the North Carolina phase of this project. Nathan Golub provided distinctive cover artwork for both volumes, fashioning a cartoon record-geek version of me that deftly conveys the flavor of these musings. I'm grateful to David Fellerath for his sage editorial advice, in person and via Track Changes—questioning, wheedling, occasionally nitpicking, frequently plugging the Minuteman, and providing much needed LOLs to let me know a line had landed, or adding a sincere "I'm convinced" at the end of one of my longish rants. Thanks, Denise Prickett, for your capacious editorial skills, which streamlined things immensely.

Thank you, Patterson Hood, for your cracking foreword, your early praise and encouragement for this project, and for all the great music.

Jeff Klingman showed up at Lotus bar one evening in the twilight of the Bush era and ended up fielding eight years of emails from me regarding my latest numerical conundrum, whether he liked it or not. Lots of other stuff too. Thanks, Jeff.

Kitty Florey—novelist, mentor, friend, and unfailing email correspondent: you were the first person to truly champion me as a writer. For that I am eternally grateful.

Ivan Malfa-Kowalski, you helped fan the flames at the bar that day. And now you are a man. Thanks, lil bro.

Thanks to Michael Horowitz—Uncle Mike—for your numerous vital contributions to my musical consciousness throughout the years, and for your valued input on this book.

Thank you, Vivian Connell, for your help with this project and for your invaluable, cherished friendship.

My hat is off to the musicians who spoke with me, whose words took my thoughts beyond mere conjecture:

Willie Alexander
Michael Bond
Bryan Bruchman
John Wesley Coleman
Anna Domino
Mitch Easter
Jared Friedman
Skylar Gudasz
Kevin Haley
Patterson Hood
Mary Lou Lord
Dan McGee
Marissa Paternoster
Ron Peno
Kim Salmon
Dwight Twilley
Dean Wells

For helpful feedback and sundry assistance, I offer thanks to Glenn Boothe, Grayson Currin, Jeremy Lange, Linda Roghaar, Bertis Downs, Jesse Jarnow, James Jackson Toth, Jefferson Hart, David Burney, Brian Howe, Keith O'Brien, Sebastian Hernandez, Kelli Douglas-Hernandez, and the Saussy family. Thanks, Jules Gray, a Twitter pal who gave this volume an extremely careful read and argued strenuously (and successfully) for the inclusion of the Beach Boys' "4th of July."

Love and gratitude to Mom and Dad for bringing me up in a household where music was always playing: Billie Holiday, Django Reinhardt, and the Beatles would have been enough, but there was much more.

Finally, I thank my boys, Nathan and Daniel, for making Dad-hood the best act yet, and to Alison: you are my happiness.

—David Klein

Notes

www.ingramcontent.com/pod-product-compliance
Lightning Source LLC
Chambersburg PA
CBHW080241170426
43192CB00014BA/2522